# THE PROFITABLE HOBBY FARM

## How to Build a Sustainable Local Foods Business

### Sarah Beth Aubrey

WILEY

Wiley Publishing, Inc.

Howell Book House
Published by Wiley Publishing, Inc., Hoboken, New Jersey

*Library of Congress Cataloging-in-Publication Data:*
Aubrey, Sarah Beth.
The profitable hobby farm : how to build a sustainable local foods business / Sarah Beth Aubrey.
    p. cm.
  Includes index.
  ISBN-13: 978-0-470-43209-9
  ISBN-10: 0-470-43209-8

    1. Farms, Small. 2. Part-time farming. 3. Sustainable agriculture. 4. Food industry and trade. I. Title.
    HD1476.A3A93 2010
    630.68–dc22                                                    2009034006

Printed in the United States of America

10  9  8  7  6  5  4  3  2  1

Edited by Beth Adelman
Book design by Lissa Auciello-Brogan
Book production by Wiley Publishing, Inc. Composition Services

# Contents

*I fondly dedicate this work to my grandma,*
*Dorothy Willard.*

# Acknowledgments

This book could not have been possible without the introduction to Wiley Publishing from Megan Saur (fellow Artist's Vineyard wine drinker in Noblesville, Indiana), who works in the Indianapolis human resources department for Wiley. From there, I was met with generous enthusiasm by Wiley acquisitions editor Pam Mourouzis, who graciously listened as I pitched my idea and delighted me when she asked for a formal proposal.

Others instrumental in the development of this work include editor and motivator extraordinaire Nancy Baxter of Carmel, Indiana, as well as each family profiled here, who warmly offered their stories of beginning their own brand of local food or agricultural enterprise. I'd also like to thank Bruce Rector of Aah Winery for some California connections, and Jerry Nelson at Purdue University for providing access to data used in the early sections of this book. An extra special thanks goes to my research assistant, Lauren Hobbs, my neighbor and Purdue University student, for her work in developing resources and generally compiling data, as well as helping organize me during the busy time as I finished writing the first draft of the chapters.

As with each work I've had the pleasure to write, my thanks must always be completed by giving his due to my handsome and loyal husband, Cary. He has been with me every step of my writing career and has supported me without complaint as I have followed my dreams.

Oh, and there is just one more (heaven forbid I forget!): This one's for you, Ace. Here you go, you've gotten "the nod" in a book!

# Introduction

I am passionate about bridging the producer-consumer gap. My purpose in writing this book, and for much of the writing and speaking I do, is to educate consumers about agriculture, cultivating local foods, and farming on scales both large and small. I also spend time educating producers about meeting customers' needs and desires and marketing directly to consumers.

I hope what emerges is consensus- and business-building for both sides. I believe that food producers have to respond (with attention, not irritation) to consumer tastes, and honestly, I believe that every consumer ultimately needs not only to understand where their food comes from, but also to have a measure of respect for the hand that produced it.

I am generally pleased to see that many Americans are interested in knowing the origins of what they eat and that many people want a choice in how they purchase foods and other agricultural goods. I am concerned that some groups are oversimplifying this choice and trying to create consumer misconceptions about one segment of the food-producing population or another.

There should not be a rift between traditional and alternative forms of agriculture. If all of us farmers are of the mind that we want to farm or create food or live a rural lifestyle, then we have more in common than we may at first realize. These goals are shared by people who want to start niche ventures in food and agriculture, whether they are lifelong farmers like me or simply long for a different life.

There seems to be a prevailing attitude that townie = liberal and farmer = conservative. These stereotypes may sometimes be true, but they aren't always. And if the goals of both groups are the same—that is, producing a wholesome product with good intentions, earning an honest living, and providing for a family—then we should not be at odds. Rather, we should enjoy the support of one another and the lifestyles we create for ourselves.

Everyone has to start somewhere.

Though I grew up as a farm kid and still see myself as that little girl in my heart of hearts, I've worked in both agriculture and consumer-oriented businesses before "retiring" from corporate life at 29 to become a full-time author and cattle raiser.

As a financial advisor (read: stockbroker with a modern, less stodgy-sounding title provided for my use), I worked directly with a variety of consumers. Money, like food, is a common need for all of us. I had women clients who were curious about the pictures on my desk of my husband, Cary, and me showing cows. Though my boss at the time scoffed at the photos as "totally unprofessional," they seemed to endear me to many clients, and eventually, that aspect happily helped free me from office life. You see, those clients realized I was a farmer and started asking me questions like:

"Do you eat those cows after you show them?"

"Do you love your animals?"

"Can we visit your farm and bring our kids?"

"Can we buy meat from you?"

They also began to make what I thought were rather amusing statements like:

> "Well, I'm a vegetarian, but I *might* eat meat if it was from *your* farm."

> "Do you actually like living in the country?"

> "You don't *look* like a farmer."

So what did I eventually do? The only honorable thing a farm girl who was dying to be an entrepreneur could: I quit, shocking my husband, and went from a good living to a start-up meats business.

Ultimately, it was through the creation of Aubrey's Natural Meats, LLC, that Cary and I began to educate urban and suburban consumers. The first thing I learned was that listening to them and explaining clearly and sincerely what we did on the farm not only sold more meat, but also created a loyal customer base almost immediately.

That happened fast, but in the last five years more realizations have emerged from the connection I created between farm and plate. I began to have conversations not so much about our product, but about how we raised the product, how we liked living in the country, and what it was like to sleep on a quiet road in total darkness. Now, if you live in the country and raise livestock, that seems ordinary; but if you don't, my life may be a novelty to you—one that you may want!

Yes, Cary and I began to see earnest desire in the faces of some customers. Yes, there are many, many people who, for very different reasons, want to make the transition from consumer to producer. If you are one of those people, you're reading the right book. I have a farm background and continue to live on my farm today, but I realize the challenges are different for someone who has come to this lifestyle from another set of circumstances. So in each chapter I profile nonfarm folks who have taken steps to establish a local food business or small hobby farm *with no background at all in farming or even in rural life*. These successful individuals and families will inspire, encourage, and educate you. (If you'd like to read more of my essays or my blog, go to prosperityagresources.wordpress.com.)

## Inspiration

Perhaps the best example of a townie turned farmer in my life is my grandma, Dorothy Willard of Rossville, Illinois—the area where I was born and raised and where the vast majority of my farming family still lives.

*Townie,* a not-necessarily-negative term that, by my definition, means "a person of nonfarm heritage," is a word I would never have dreamed applied to Grandma. She was my image of the perfect farm woman of her era.

Having been raised to believe that farming men and women were heroes and that friends from town were "not quite the same," I have to admit, when I found out that Grandma had been raised in town and had not a clue about farming when she married my grandpa, Ray Willard, I was absolutely horrified. This truth, though I certainly wasn't willing to believe it at first, was far worse than finding out that Santa Claus is a myth. Really.

Grandma was farm life embodied in a person. And in the soft, hazy memories of my youth, I picture us in a light much like the blurred edges of daytime during a humid Midwestern summer as we went about all things rural. I planted, weeded, and harvested the garden with Grandma, she in her massive straw hat and bandana. When I visited, every morning we took scraps out to the sows, who grunted and ambled toward the fence and Grandma's billowing nightgown as we tossed the greedy porcines potato peels and coffee grounds.

Along with my cousins, we walked through the wheat field that had been cut for straw back to the creek that ran shallow but cool on hot days and picked cherries, peaches, and walnuts from her orchard. We chatted enthusiastically about my 4-H projects in hogs and cattle, and she knew everything about taking care of wayward barn kitties and even shot at the occasional honking goose that caused a disruption. To this day (when this book is published, Grandma will be 84 years old), she is comfortably conversant and knowledgeable enough to talk with me about our natural meats business, our show cattle business, and commodity prices for corn and beans. Townie? Surely not!

Still, truth was, in 1946 when Grandpa returned from World War II and they were married, Grandma said she couldn't even light the stove to boil water. Grandma not know how to cook? She's an amazing cook! I didn't believe it.

Yes, Grandma told me, she'd been raised the only child of an aspiring artist and poet mother and a local entrepreneur father (he owned a grain elevator). Although the family was not wealthy, Grandma was raised with luxuries I never would have thought of as normal, such as a laundress. Before she got married and moved into the brick home her new in-laws had vacated so that Grandpa could take over the farm, Grandma had never cooked, nor had she performed the litany of tasks I always attributed to her with a kind of pioneer heroine worship. She didn't know how to garden or preserve every known vegetable in the garden; she'd never picked and pitted cherries to make preserves and pie; she'd never come near a hog or driven a tractor.

In later years, I often wondered just how she did it—that is, convert herself into a bona fide country woman. Volunteering with the pork producer's association, balancing a farm checkbook, leading a 4-H club, and hosting the local Ladies Home Extension (which she still does) must have been far from her mind as a young woman of 20. Yet Grandma made the transition from town to tractor (though not seamlessly, she says) and raised all her children and grandchildren to grow up on farms and in the rural lifestyle.

My grandma is the most practical and pragmatic person I know. I asked her once how she always managed to look at things objectively and how she seemed to take disappointment and irritation so well. She laughed, and then, when pressed, she shrugged. "You know, Sarah Beth, what are you really going to do about it, anyway? You've just got to go about your business and enjoy life."

**A rather simple recipe—**but if it keeps a smile on your face for 84 years, it's a rather useful one at that.

You can probably tell that I think a great deal of my grandma. She also attempted to teach me to sew, took my side when I fought with my parents as a teenager, and still shares my love of good books and crossword puzzles. It's because of Grandma that I was inspired to write this book for you, the aspiring small farmer or local-foods business owner. Whether you want to create a small business or just farm for the pleasure of it, this is the book for you. If Grandma could do it, so can you!

## How to Use This Book

Maybe you've come to your hobby-farm idea by default, maybe you've been pondering it for a long time, and maybe you've just ended up in a place where you're looking for something meaningful to do. No matter; you'll find a fellow and maybe even a friend in any one of the profiles in this book's pages.

While you'll find much inspiration, motivation, a bit of editorializing, and lots of how-to, this book is meant to genuinely help you start something new. To accomplish this objective, I've organized it so that you should be able to look for a topic in the table of contents and find a section that covers it.

At the beginning of each chapter is a list of learning objectives to help you gauge the scope of what's covered. Each chapter contains profiles of other nonfarmers who've made a life for themselves and their families as farmers. Also in each chapter, I offer my best tips and ideas. And finally, you'll find a box with a list of best practices—wisdom and experiences from the entrepreneurs profiled that are relevant to the specific topic at hand. The box encapsulates their wisdom into a quick checklist for you to refer to again and again.

Having worked for a time with corporate training in the agricultural business sector, I learned the importance of letting adults know in advance what they can expect to learn along the way. So, in that spirit, I offer the following learning objectives.

By the end of this book, you should be able to:

- Understand the trends in niche agriculture and local foods
- Create a business plan for a small-scale or hobby venture
- Assess your home marketplace and uncover niches not yet served
- Learn about financing options, especially grants and cost-sharing programs
- Learn how to start a CSA (community-supported agriculture) or a food cooperative
- Learn how to start an agritourism business
- Identify and implement free or low-cost marketing ideas
- Choose from a variety of selling venues for your products
- Pick up tips on getting chefs and other local businesses interested in your products
- Create, administer, and grow an e-mail customer database and send newsletters and correspondence that generate sales
- Understand basic strategies for surviving the first year in business

Because there are different stages in the entrepreneurial process, the resources section will be very useful when you are seeking information or ideas. You'll find contact information for organizations ranging from financial groups to trade associations. Contact information for every person profiled in the book is included as well. Finally, you'll find sample forms and documents in the chapters and in the appendix. I've used every form here, and the template has worked for me.

## Sarah's Rural Best Practices

Friend to friend, I offer for you a few country commonalities that you just might want to know. (And by the way, there's no such thing as cow tipping or snipe hunting.)

- Rural people still drop by unannounced and want to chat. While they don't necessarily plan to come into the house, they probably do expect you to at least *offer* them iced tea or a beer.

- Drive slowly on rural roads. It's rude to fly by someone's house kicking up gravel and dust.

- In the same way that you wouldn't pass a stopped school bus carelessly, give slow-moving farm vehicles space and courtesy.

- A person you don't know who waves with one or two fingers is not giving you an internationally impolite sign. It's more than likely a "farmer wave," and you're welcome to return it.

- Neighborhood dogs may wander. Don't be alarmed unless they appear unfriendly. It's also nice to call around and see who's missing one.

- Coyote and deer at night are common, especially during grain harvest and hunting seasons, so keep your eyes peeled.

- Cows are usually friendly, except when they have little calves.

- Yes, pigs and horses can bite; ask first before petting.

- Sheep do tend to scatter and follow at the slightest whim.

- Goats will eat anything.

- If it's a wire fence, it's probably hot (electrified). If you must find out for yourself, it won't kill you, but you won't forget it, either.

- The funky smell in the air is the smell of money to someone, and that someone was there before you.

# Part One

# Creation and Veraison

# Chapter 1

# The Case for Small-Scale Farming and Local Foods Ventures

---

**Learning Objectives**

🌱 Understand some facts about the number of new small agriculture and food ventures and who's starting them.

🌱 Learn about organizations that are supporting these types of start-ups.

🌱 Review data on the size of the market for new local foods businesses.

🌱 Meet the unlikely Lavender Queen, Jeannie Ralston.

🌱 Discover compelling personal reasons to move from town to country.

---

## Creation and Veraison

The first half of this book is devoted to the process of going from an idea to a real business plan and actually getting started. It's a time of critical change, which is why I call part one "Creation and Veraison."

What do these terms mean in the context of this book? Of course, creation is the genuine start of something from the root or the source. Creation can imply something wholly spiritual or as concrete as production and manufacturing. Creation can involve one person's ideas or a twosome, such as pollination in plants. No matter what images come to mind, when you read the word, the notion of beginning is always there.

The first half of the book covers that initial impetus, but also the evolution of a little idea into a real action. That's why the word *veraison* is also appropriate. Veraison is a wine-making term that refers to the critical time when the green grapes begin to change to their true varietal color. It's the stage at which most of the ripening takes place—when the sugars begin to materialize and the grape comes into its proverbial own—and is critical for the wine.

For your business and hobby endeavors in niche agriculture, your personal veraison is the early stage, beginning somewhere between idea and start-up. Your veraison occurs when you truly begin the long process of becoming something new.

Every one of us arrives at the precipice of creation and veraison from a different point of view and with different goals. What we share is that we are part of a growing trend. After three or four consecutive generations of leaving the farm and the kitchen in search of convenience and a faster-paced lifestyle, people are coming back to agriculture.

What is driving this trend, and where is it headed? Some statistics claim organic farming alone is doubling every year or so. So is the market for local foods and niche agriculture already flooded, or is there room for you when you get here? How difficult is it to move from urban or suburban to rural, and how different is the lifestyle? Can you really make a new business work?

## What Are the Trends?

In this chapter, I'll examine some of the trends and the size of the marketplace in small farming and local foods. I'll also introduce a few organizations and associations that support these trendsetters through membership, promotion, and education. (For a longer list of organizations and groups to contact or research on your own, see the Resources section in this book.)

The list of niche agriculture and food ventures that nouveau foodies and farmers are creating is nearly endless, and is as boundless as the imaginations of the proprietors. Many niche businesses are marketed by appealing to consumers with an interest in foods labeled as natural, organic, or free-range. In fact, that segment of the food industry is about to outgrow the term *niche*. And the market is not anywhere near its zenith.

In the chart on pages 5 and 6, I've listed some common types. Clearly, there is a broad diversity of niches and business ideas. In this book, you'll meet producers and urban foodies who have started hobbies and businesses in all of these areas and more.

A field of lavender was one woman's entry into the world of small-scale farming.

Some of the terms in the chart are farming ideas that are also marketing ideas, and are commonly combined together to best meet the target customer's wishes and needs. Packaging these features can seem daunting. And if it's not based on a solid foundation and done well, marketing in small-scale agriculture can confuse the customers. Using examples of peers who have created a small farm or food business, I'll not only excite and motivate you, but also show you how to effectively bundle together these niches to create a profitable business.

Before I go too far into the facts and figures, let's meet someone who epitomizes that journey from townie to farmer in such an interesting way that she wrote a book about it. Her personal story makes a case for starting a food or hobby farm business better than any other tale I've heard. And her journey to starting a rurally based lavender business shows that no matter how you arrive in the country, there's a business idea waiting for you.

## Niche Markets in Local Agriculture

| Agriculture and Animal Production | Cooperatives and Community Organizations | Local Businesses That Support Small Farms |
|---|---|---|
| All-natural (including no hormones, no antibiotics, no preservatives) | CSAs (community supported agriculture, a community of individuals who pledge support to a farm operation so that the farmland becomes a kind of community farm with shared risks and benefits) | Local delivery services (delivering local or organic farmers' goods) |
| Organic | Food cooperatives | City markets and farmers markets |
| Grass-fed | Buying groups or consumer buying cooperatives | Retail shops that stock local foods |
| Breed-specific (such as Berkshire hogs or Belted Galloway cattle) | | Artisan cheeses, soaps, and handmade foods made with local ingredients |
| Free-range | | Wine making using local fruit or purchased grapes |
| Nutrient-enhanced (such as adding omega-3 fatty acids or vitamin E in the feedstuffs) | | |
| Locally raised | | |
| Insecticide-free or pesticide-free | | |

(continued)

| Agriculture and Animal Production | Cooperatives and Community Organizations | Local Businesses That Support Small Farms |
|---|---|---|
| Bird friendly, predator friendly, or fish safe (among numerous other ways of indicating that no wild animals have been harmed in producing these products) | | |
| Sustainably or biodynamically farmed (grown in a way that views the farm as a self-sustaining organism) | | |

# Jeannie Ralston, an Unlikely Small Farmer

I spent a fair amount of my time chuckling at Jeannie Ralston's lack of understanding of the rural lifestyle as I read her 2008 book, *The Unlikely Lavender Queen: A Memoir of Unexpected Blossoming*. It's a sometimes funny, sometimes bitterly personal, and sometimes painful story. When Jeannie moved to Texas from New York City to follow her new husband's ambitions, it was obvious she had no clue about country living.

She was still trying to adjust to country life and motherhood when her spouse insisted they start a lavender farm. Jeannie is married to Robb Kendrick, an internationally acclaimed photographer who travels extensively for his work. Although lavender was his interest, not hers, she knew she would end up running the farm. "I have an immense respect for people who make their living off the land, but farming was so foreign to me it threatened the image I had of myself," she wrote.

In her book, Jeannie recounts her initial disdain at planting lavender, harvesting, and then marketing the flowers to local stores (not to mention her limited tolerance of the oppressive Texas heat). The more I read, the more I saw her as the ultimate townie—the antithesis of me. And that's exactly why I wanted so much to interview her for this book.

Jeannie went from city to country kicking, screaming, and crying, and yet found out that she loved the life and loved her rural small business. Jeannie discovered things the hard way and learned a lot of lessons through trial and error.

Like me, she is both a small farmer and an author. What we don't have in common is how we arrived at our place in the world. I'm a farm girl

many generations back, looking to carve out my own small farming niche while writing for the likes of *Country Woman, Beef Today,* and *Farmworld.* Jeannie, a city slicker, writes for the *New York Times, Allure, Parents,* and *National Geographic.*

Jeannie's story is compelling, but the real reason I asked her to be part of this book's first chapter is because it is not entirely unique. If you're reading this book, you've got at least something in common with Jeannie, whether it's a strong desire to begin a hobby farm or a strong desire to find a good use for your energies while living in the country. Either way, the migration to local foods and niche agriculture is growing faster than anyone predicted.

Jeannie Ralston, her husband, Robb Kendrick, and their children.

# How Big Is This Market?

After generations of leaving the country for cities and suburbs, Americans are coming back to the country—or at least to a rural-esque lifestyle. And they are seeking something more than to simply live in the country; many families want to experience a bit of the rural life by raising small flocks of animals or growing large vegetable gardens in town. These individuals are starting small local foods businesses, farmer and consumer cooperatives, and farm-fresh delivery services. Mail order and Internet-based food businesses and new farmers markets are appearing every year.

## Farmers Markets

Since 1994, the United States Department of Agriculture (USDA) has been tracking farmers markets and publishing a directory every two years called *The National Directory of Farmers Markets.* Growth in the number of markets

A group of shoppers enjoys a fall farmers market. The number of farmers markets is increasing just about everywhere.

has not been explosive, but it has been steady. While many commercial sectors seem to expand and contract through various economic conditions and the political parties in office, the number of farmers markets in the United States continues to grow every year.

From 2006 to 2008, the USDA reports that the number of markets increased by 6.8 percent, from 4,385 in 2006 to 4,685 by August 2008. That means that since 1994, when the USDA officially began attempting to keep track of market numbers, more than 3,000 farmers markets have been added across the United States. (I say "attempting" because markets change so often, and while the USDA encourages participation in the surveys it uses to keep the directory up-to-date, there is no requirement to participate, nor is there a requirement to register with the USDA to open a farmers market. I personally know of several markets that aren't included in the USDA's directory, even though the USDA works with state departments of agriculture and farmers market associations at the local level to collect information.)

## Food and Agricultural Cooperatives

The concept of a cooperative has been used in rural communities for millennia, as farmers and citizens of small towns helped one another with everything from protection to harvesting to medical care. While this kind of community has all but faded into history, food cooperatives and community-supported agriculture (CSA) are reversing that trend.

CSAs are a type of cooperative. They're usually formed by one farmer or a group of farmers who then sell shares in the CSA to the public, making those shareholders members of the cooperative. Members of the CSA receive food items and agricultural goods regularly; many go out to the farm to pick up their produce, herbs, meats, and other goods every week during the summer.

Sources vary about the exact date, but in the mid-1980s the concept of the CSA was brought to the United States from Europe, where these organizations had been growing for almost 20 years. Growth here has been steady, as with farmers markets. The National Sustainable Agriculture Information Service reports that there are almost 1,100 CSAs; other trade groups record a number closer to 1,500.

Expansion has been rapid since 1999. While CSAs are scattered across the nation, the burgeoning hotbeds seems to be on the two coasts and in the upper Midwest, in states such as Wisconsin. These areas, according to the Robyn Van El Center (a nonprofit educational center that supports the development of CSAs), each have between 4 and 10 percent of the nation's CSAs. Geographical and size data for CSAs in the interior of the continent seems a bit hit-or-miss, so estimates are really just that. Additionally, an estimated 10 percent of CSAs are organized as nonprofit groups. I believe unprecedented growth in this niche is coming soon as consumers seek ways to become involved in small farming and as the interest in locally raised food continues to expand. That's why I have devoted chapter 4 to exploring this local foods and sustainable agriculture model. I've also assembled a large group of resources in the appendix that will provide you with contact information for many of the groups listed throughout the book.

## Natural and Organic Foods

Any book about opportunities in local foods and small agriculture would be incomplete without a discussion of how natural and organic products have fueled the growth of all market segments. In fact, I could write a whole book about just these two segments—another time, perhaps.

When talking about food trends, we need to start by understanding the difference between *natural* and *organic*. In 2001, statistics from the USDA Certified Organic program showed that up to 75 percent of Americans surveyed thought *natural* and *organic* were the same things. They are not. The simplest way to distinguish them is that food labeled as USDA Certified Organic must meet the standards set forth in 2002 by the Organic Foods Production Act. These include a stringent set of criteria for feedstuffs and other production practices.

## How Does a New York City Writer Become the Texas Lavender Queen?

"This is all Jeannie Ralston's fault." That's what the residents of Blanco, Texas, say about the increased taxes and traffic, and Jeannie laughs when she repeats the remark. Blanco is where she and her husband, Robb Kendrick, founded Hill Country Lavender in 1999, and it's true that a steady growth in property values and tourism has bloomed in what is now the official Lavender Capital of Texas (a distinction Jeannie fought hard to obtain from the state legislature). Blanco has become an agritourism destination based around lavender.

Jeannie also helped start the annual Lavender Festival that draws thousands of tourists and lavender growers each June. And she hosts workshops to teach budding entrepreneurs the tricks of the heavenly scented trade. "It's not just about me and my husband," she says. "The town saw this as an opportunity and used lavender as a part of its image; they just embraced it."

While she found her rhythm in the undulating purple haze of her lavender fields, Jeannie didn't start out to become the Lavender Queen. Her business started out as a kind of obsession of her husband's. The family spent some time in Provence, France, where Robb was photographing for *National Geographic.* Jeannie loved Provence and the fields of lavender, calling them a "mélange of Van Gogh colors." But it was purely an abstract interest. Robb, however, had other ideas. He spent his free time in Provence asking a local farmer about how lavender is grown. When they all returned to

*Natural* is a much broader term. It is usually not governed by any specific labeling requirements, though Country of Origin Labeling (COOL) regulations, which went into effect in March 2009, may change that. COOL makes labeling the country of origin mandatory on most fresh meats, including beef, chicken, pork, goat, and lamb, as well as on other perishable agricultural commodities such as produce. Because labels are now required for these products, there may be additional requirements for label claims as this legislation gets implemented over time. (For more information on COOL, check out www.countryoforiginlabel.org.)

### Market Growth

Once considered niche markets, natural and organic products have gone mainstream. In 2007, the organic market yielded a startling $20 billion in annual sales, up from less than $1 billion in 1990, according to the Organic

Texas, he put out his first few plants and tested the hardiness of various varieties in the local climate.

By the second summer of trials and plantings, Hill Country Lavender was in business. Jeannie was the proprietor who hosted people on weekends to pick blooms right off the plants. Eventually, she began adding a variety of products to complement the lavender bouquets, and suddenly she and Robb were conducting workshops. When summer arrived each year, Jeannie found herself busy—and liking it. When a tourist cooed about how wonderful her life must be living on a lavender farm, Jeannie recalls in her book, "Though it was true I had never had such a dream, I was beginning to see why others might."

However, she resists the notion of "the heroine moves to the country and everything is perfect." Making the transition was difficult for her, required a lot of work, and wasn't the first choice of careers for a successful freelance journalist. Still, lavender and her small farm played an important role in bringing Jeannie back from a bout of depression and certainly gave her something to write about.

It took years, but Jeannie's lavender business became the dream she never planned. "I had endured, toughed out the isolation, the demands of a perfectionist husband and had found real peace," she wrote. "I felt that, like the lavender, I was a nonnative transplant that had somehow thrived."

Trade Association (OTA). These numbers exploded to $24.6 billion by the end of 2008. And that's just the beginning. The Agricultural Marketing Resource Center (AgMRC) estimates that the natural and organic market is set to double *at least* every three to five years.

Organic and natural products encompass everything from tissues to coffee, but it's food sales that consumers are driving with their dollars. Organics are the fastest-growing sector of the food and beverage industry. Yet, according to the OTA's web site, the nonfood segment represented $2 billion in annual sales by the end of 2008. And it is the nonfood segment that seems to be growing the fastest; sales of nonfood organics grew by 26 percent in 2006 over the previous year, while organic food products saw an aggressive 20.9 percent growth. All in all, the OTA reports that in 2008, 69 percent of adults bought organic products at least occasionally, with 28 percent buying weekly.

Out here in the country, we know from the number of interested consumers that producing natural and organic products makes money, and our market is in no way saturated. Sure, the mega retailers like Whole Foods Market are pocketing most of the consumers' dollars, but at the local level there is room to grow because consumers want to buy from companies and people they know.

**On my own farm** in central Indiana, we entered the natural products market in late 2003 with the formation of Aubrey's Natural Meats, producers and purveyors of beef and pork raised without added hormones or the use of antibiotics. Our animals are also raised outside on pastures. Everything is raised locally; for us that means in our home county.

### Why the Demand?

Many producers want to know why consumers have such a strong interest in organic products when our traditional food system has, in their view, served the customer very well for generations. This is a touchy subject with farmers, with eaters, and with anyone who wants to combine both into a lifestyle.

The reasons are varied, and I believe that individuals make their food choices for very *individual* reasons—some serious, others less so. Some reasons are:

- Perceived increase in food safety
- Perceived health benefits
- Allergies to certain chemicals
- Desire not to consume pesticides, insecticides, or added hormones
- Desire to better the environment
- Desire to support small farmers
- Preference for the quality or uniqueness of some organic products that are not offered in any other form

The message underlying all these reasons is really important for farmers to hear: Consumers want choice, and they want producers to listen to their needs.

## The Local Foods/Slow Foods Movement

The average food item travels 1,500 miles before an American consumes it. That's a long trip! So is local the new natural? It's a question I recently attempted to answer for myself and the readers of the column I lovingly pen for Food Trends, a section of *Indianapolis Dine* magazine, where I try to draw parallels from farmer to foodie.

But what does *local* actually mean? Well, that's as relative as defining the term *natural*. Some would say local means foods produced within a certain regional radius of the consumer—say 50 to 150 miles. Others describe local foods as those produced within a tri-state area. Still others consider local eating to be more micro than that, breaking it down to eating only what is in season and can be obtained fresh locally.

So which has the bigger market: naturally raised foods or foods the consumer buys locally and eats seasonally? Are they related or mutually exclusive? Is one more important than the other? Which is better?

It's actually becoming a bone of contention as a growing number of educated consumers argue that eating local is the most sustainable and least expensive model and yields the smallest carbon footprint. The debate has been fueled by the book *The 100-Mile Diet: A Year of Local Eating,* written by Canadian couple Alisa Smith and J. B. MacKinnon, and others like it.

Small-scale local eating has become a huge market. Packaged Foods, a food market research group, estimates that the market for local foods was valued at about $5 billion in 2007 and is expected to grow to $7 billion by 2011. (Other facts and information can be found at www.foodroutes.com and www.eatlocalchallenge.com.) One of the leading organizations that helped spark the trend is Slow Food USA. The group's name has become synonymous with the slow food movement—the lifestyle surrounding the enjoyment of healthy food while making the connection to family, community, and culture. Slow Food USA was founded in 1986 and now has 80,000 members from 100 countries. There are more than 70 local chapters in the United States. The group seeks to influence changes in food policy, production practices, and market forces so that they ensure equity, sustainability, and pleasure in the food we eat.

## Agritourism

Basically, agritourism means making your farm a vacation destination. I cover the topic in detail in chapter 6, but meanwhile, AgMRC reports the following annual values for agritourism in these states:

- California: $51.8 billion
- Colorado: $2.2. billion
- Hawaii: $33.9 billion
- Vermont: $19.5 billion

## The Economic Impact on Rural Communities

Rural economic revitalization, development, and jobs are big issues for people who live in communities with fewer than 20,000 residents. As farms have consolidated and big box stores have driven local retailers out of business, small towns and the farming communities that support them have been struggling for a generation. Local leaders worry about everything from quality of education to brain drain, when promising young people depart to find better-paying jobs in larger metropolitan areas.

One of the tremendously important aspects of the small farm and local foods movement is the impact these new twists have on jobs and income all across America. Small farms don't compete with large growers, so neither group should feel threatened. Both are needed to produce the foods and fibers that consumers want and create the jobs small towns need.

Jeannie Ralston is an example of a rural economic developer, although she certainly never thought of herself that way. It's just the natural by-product of starting a rurally based venture.

"I always felt if there was a bigger group doing something, then everybody wins and it [in this case, lavender] becomes more of an attraction," Jeannie says of her time spent organizing people to help develop Blanco, Texas, into a tourist spot.

Jeannie found that working with her neighbors created business. "The cooperation part is huge," she says. "You can't do it all on your own. Go and work with other people, other farms or crafts. Find something that represents the area. The other guy down to road is your partner, not your competitor."

Rural enterprises are creating new rows in old fields, adding millions of dollars to local economies each year. How do you start bringing the dollars to your hometown? Begin with the motivation—even before the know-how, advises Jeannie. And be willing to stick out the tough times. "It depends on the town you're in, if they're willing to embrace it. It can be a slow process. If it is something new, there may be resistance, but with diligence and patience you can help people see it can be a way to help the town."

AgMRC also estimates that agritourism has some economic impact in every state. About 2.5 percent of all farms in America receive some kind of income from hosting guests to the farm. That translates to more than 52,000

Agritourism means vacationers come to your farm, often for the day. This is the patio and deck at Ravens Glenn Winery. (You'll meet the owners in chapter 10.)

farms in America. Some states are big leaders in agritourism; in Vermont, one-third of all farms have at least some agritourism business.

# Trends in Suppliers

With the growth in demand has come a growth in supply. When it comes to supplying farm products, that generally means groups of people moving to the country. But being in the country, by itself, doesn't make a person a farmer. The trends also show an increase in the desire for education and networking to support new agriculture business ideas.

## Trade Associations and Conferences

We are a people who learn, and we love to congregate with those of like mind. Trade groups, conferences, and organizations that support new farmers and local foods host meetings that in some cases are standing room only. Membership in many new groups is also growing at a fast clip.

Much of the growth in conferences and resource materials is at the state and local level, so facts are often supplied by micro-groups such as a county cooperative extension office in a local region rather than by large, national organizations that track data across many states. Every group I investigated for this chapter reports that they are growing. Here are a few national examples.

The 2008 Organic Farming Conference, hosted by industry-leading organization Midwest Organic and Sustainable Education Service (MOSES), had 2,400 participants, up from 90 just ten years ago. Most of the farmers who go to the conference are making the transition from traditional agriculture to organic. It also attracts people who are new to farming.

The Ecological Farming Association, also known as Eco-Farm, is considered the world's largest mainstream organization committed to furthering natural, organic, biodynamic, and sustainable agriculture practices. Eco-Farm's education programs have reached more than 50,000 participants during its 26 years.

## Urban and Suburban Transplants

According to population data and real estate agents, young people, retired couples, and aspiring Generation Y entrepreneurs are all moving to the country. In October 2008, the *New York Times* reported that despite the impending recession, the financial bailout bill bedlam, fears inspired by the credit crunch, and a hotly contested election, migration to rural areas from urban communities remained strong. So strong, in fact, that it appears to have officially bucked the trend of urban migration that has prevailed in the United States since the Industrial Revolution.

In fact, this hot new trend is not really all that new. Back in 1980, the U.S. Census reported that small towns and rural areas experienced faster growth than urban and suburban areas. In the housing market, trends in favor of country living remain strong and have for nearly two decades—although there have been short-term spikes and dips. The federal government reports that housing prices in rural areas tend to appreciate more rapidly than housing prices in metropolitan areas.

The move from town to country is even more pronounced in areas that are considered high-amenity,

> I always get a kick out of it when people tell me, "You don't *look* like a farmer." I think it should be apparent by now that even I don't know what a farmer is supposed to look like. **Stereotypes be gone;** the city has gone country and country is now cool.

such as places with gorgeous natural scenery or regions that lie within 20 to 30 miles of big attractions such as theme parks and popular festivals.

Of course, not all these rural transplants are starting small-scale farms. Not every rural person is a farmer. In fact, according to the USDA, just below 2 percent of our nation's population resides on an agricultural enterprise.

How many of those farmers live on small-scale farms? It depends on how you define a small farm. So indulge me in presenting just a few more facts and figures.

As of 2007, the USDA reports there were 2,076,000 farms in the United States, with an average size of 449 acres. Despite the growth of small, hobby, and lifestyle farms, that number is down from a 1990 high of 2,145,820 farms. An acre is about the size of a football field, but while that may seem like a lot of land if you live in an urban area, it is not a lot of land in terms of commercial-scale crop or animal production. At 449 acres, a farm of that size would be considered medium to small by aggie standards—though for producing locally raised foods sold direct to consumers, it would be quite large. The size is all in how you look at it.

The USDA's agricultural marketing service reports that 94 percent of all farms are small farms—that is, operations with gross receipts of less than $250,000 per year. (Remember, gross receipts are the money you take in before you deduct the costs of doing business. It doesn't mean most small farmers make that much in profits.)

A whopping 40 percent of all farms are considered lifestyle or recreational, with some 834,000 plots of land dedicated to hobbyists. I should note, however, that these numbers seem somewhat inflated to me because the official USDA definition of a farmer is an "agricultural producer," and to be considered an agricultural producer one need only market $1,000 or more of agricultural products per year. So it doesn't take much to get started, at least as, shall we say, a quasi-farmer.

## The Human Reasons for All This Growth

I could cite lots more facts and figures on the growth in both demand and supply for small farms and local foods businesses. But I prefer to leave the hard data collection to web site content managers and the cooperative extension service. It's not that the data isn't important. It's just that, for me, numbers are not enough to make the case for starting a new business venture or even taking up a serious hobby. I am a writer, after all, not an economist, so streams of statistics impress my sentimental mind less than what I hear when I listen to people talk and look at the expression in their eyes.

So why are people coming back to the country and taking up farming? And why are they finding plenty of customers to buy what they produce? Let's look first at what's driving the demand.

## Who Is Driving the Demand?

When I began getting serious about putting together a demographic picture of my customer base, I found myself almost at a loss. I discovered that my customers could be broken out into several categories.

- **Sentimentalists:** They grew up around agriculture and miss it, or are one or more generations removed but had a relative in agriculture.

- **Foodies/gourmets:** They love all things delectable, delicious, rare, haute cuisine, or otherwise fancy when it comes to foods and other products.

- **Environmentally conscious:** They believe buying local is especially important for reducing the effects of large-scale production, transportation, and distribution on the environment.

- **Health conscious:** They have food allergies or believe they will derive health benefits from eating local or nontraditionally grown foods.

- **Bulk buyers/price shoppers:** They buy in volume—such as half a side of beef, which wouldn't be carried in any store—to save money.

- **Vegetarian converts:** They became vegetarians because they did not like the way meat animals are raised or felt modern meat production methods made meat unhealthy; free-range and other less intensive animal husbandry techniques have won them back.

It's a broad group, I know. As you might guess, many of these categories overlap. But they all exist. And to be honest, at first I was perplexed that I couldn't come up with one or two categories into which I could stuff all my customers, allowing me to craft what I imagined were market-savvy customized messages. But I can't.

I spent two summers evaluating the data I gathered from voluntary customer surveys that led me to these groups. Finally, one market morning as the summer waned and pumpkins replaced sweet corn as the wait-in-line-for-an-hour produce of choice, I finally got it. The marketplace answer I'd been looking for was staring me in the face: *This trend toward smaller farms and local foods is drawing very different types of people together.* Their backgrounds are as diverse as the reasons my customers buy from us. It's a simple and yet profound insight.

## Who Is Meeting the Demand?

As I've already mentioned, moving to the country has been a steady demographic trend. In addition, all sorts of people who already live in rural areas, or in the suburbs and even cities, are turning to small-scale farming as a way to make or supplement their income.

Jeannie shared some ideas with me about why people are so attracted to the farming life. "First of all, it's the beauty and the peace," she says. "When you're in the city, you never really see the stars."

I wouldn't argue with that. There is nothing more pristine than a cold December night when my husband and I head outside at about midnight to check on the mama cows and their babies. The ground is lit by the shimmering stars set in their deep navy blue canopy. Yes, the stars are certainly one of the big advantages.

I wondered if there are more men than women driving this lifestyle revolution. Jeannie doesn't think so. "I think it's equally split; both men and women have a desire to move out." However, their reasons for doing so may be different. "The propelling force for men seems to be that they're tired of high pressure and have a desire for physical work. It's the idea of playing with tractors. Women seem to like being outside and having more space where they can have fresh air and get the kids outside and maybe have a nice garden."

Of course, these are generalizations, and it can be hard to generalize about decisions that are so personal. But Jeannie did say that she sees several broad groups of people in her lavender seminars.

A winter's dawn seen from my own back door.

## Jeannie's Personal Transition

"If you're used to going to an office every day, this can be a big transition," Jeannie cautions newcomers to the small farm lifestyle.

"There is great romanticism about the country life that's shown in movies and written about in books, as if it's ideal. But things are not that simple. You're rather away from things (like shopping, restaurants, and services), so you're probably expected to do more for yourself." Jeannie remembers the time she was home alone when the water pump went dry—a big deal in Texas!

"Plus, I think there are things to deal with internally," she continues. "You've got to be able to be by yourself, and you might have to deal with feelings of isolation. These are good things to learn, but it's a transition. There are accommodations you have to make."

Jeannie urges other people who are thinking of starting a small farm business to really make sure they are ready to change their lives. "Ask yourself, are you an adaptable person? You'll have to get settled in; it won't happen right away."

Jeannie also urges country newcomers to consider that moving from one set of local norms to another, even if it's only a few miles, can be considered a switch of cultures—and that's never easy. "In the beginning, I was very judgmental," she admits. "I would say that if you travel outside the U.S., you'll understand that it is a separate culture and respect that." She advises new farmers to treat smaller communities and farming areas the same way they'd treat any new culture. "This all existed before you came here; respect people's ways of doing things."

As for Jeannie, "I'm more comfortable now in both places. I could live in the country or the city now. This feeling is very rewarding for me. When I was in New York City, I just thought of myself as a city girl. I was limiting myself to this one type of world, and it was the only one that was valid for me. But knowing you have the confidence to be self-reliant, that comes from living in the country."

Retirees make up what she thought would be the largest demographic moving out to the country. "They want to keep their days full and interesting and new, and I think they also want to have something productive to occupy themselves," she says.

Jeannie was also initially surprised to see so many younger people, in their child-rearing years, moving to the country and wanting to start more of a vocation than an avocation. Many young families desire the lifestyle that rural living offers but realize early on that they can add income that will supplement their main earnings. "People don't want to feel put out to pasture, especially those in their working years, and they don't want to feel like they are retired or wonder what they're going to do in the country," she explains.

She also saw a certain number of country converts who wanted to resurrect a fond memory of Grandpa's farm or summers visiting friends on the farm.

We both agreed that these very human motivations are driving a genuine trend, not simply a fad. "One thing for me is just how many people continue to be interested in our lavender-growing seminars; I'm amazed at how they fill up," says Jeannie. "Most of these people are from the city or at least a suburban area—that's how I quantify the continued interest."

## Making the Transition

As cities sprawl out into the country, overtaking farmland, there is a natural curiosity among the urban urbane to experience the rural. But things don't always go smoothly. "When suburban meets rural, it creates conflict," says Jeannie. "Blanco itself [is now an agritourism area] and it has really changed. In Texas there are fewer huge ranches and more smaller plots that people can buy and do small-scale farming."

Jeannie, Robb, and the kids in their lavender patch.

While the influx of city folks might make some farmers and ranchers uncomfortable, it works the other way, too. Moving to the country is not always idyllic.

In her book, Jeannie writes about her first experiences in rural Texas. "As we drove into the town of Blanco and saw the courthouse and the Bowling Club Café with the convention of pickups parked out front, I was even more depressed." Yet just a page later, her own veraison was already beginning: "You couldn't have told me then that I would soon feel

## Planting Seeds for the Future

In 2006, Jeannie sold her lavender farm to a young woman named Tasha who had worked for her. But the sweet-scented herb is still part of her life. Jeannie continues to be involved in the Lavender Festival in Blanco and offers seminars on lavender growing.

In 2008, she also launched the Seed Campaign. "If people buy my book via my web site, www.jeannieralston.com, the money from the commissions I make on its sale will go to a variety of charities," she says. She adds that what she makes from those sales is "found money" and "it just feels better to give it back."

the exact opposite, and I would never have believed the part lavender would play in my about-face."

# Summing Up My Case for Small Farms

In the end, why do people want to buy products from small farmers? I'm inclined to believe there are powerful socioeconomic ideas at work that have significant weight when we look at consumer trends. For example, many consumers believe it is important not only to know where food comes from but also to reassure themselves that it is wholesome. This is a serious message they are sending to federal regulatory agencies and also to lifelong agricultural producers.

There is also a lighter side to this trend. Could it be that we've decided shopping for food—at least for local, gourmet, or otherwise interesting foods—is fun? I think the foodie trend is a hobby that we willingly spend our "extra" money to feed. I have to say, if I have to choose between browsing a farmers market on a breezy morning or slogging through a big box store with everyone else in a 50-mile radius, my pick is obvious.

There may be something, too, in our desire to reach out to other human beings in a high-tech, otherwise increasingly impersonal world. Is there something in the hearts of suburbanites that tells them that at least once their children should shake the hand of the person who literally made their dinner? Is this movement to small farms and local foods getting us back to our roots? I think so, and I'm all for it.

It's important to realize that traditional agriculture is not bad or evil; we need large-scale production to *immediately* feed a huge and hungry world.

But small-scale farms, tended by devoted and loving farmers who have a strong desire to make this their lifestyle, should be able to feed those farmers, their communities, and those customers who want a choice.

## Jeannie's Best Practices

- Learn the rural etiquette from your neighbors and the community.
- Be understanding and respectful of cultural differences.
- Get involved locally.
- Start a new venture as supplemental income, not as your entire cash flow.
- Enjoy the lifestyle as well as the business opportunity.
- Don't be prissy; be game for adventure!

# Chapter 2

# From Idea to Inventory: Planning and Assessing the Market

**Learning Objectives**

- Meet Brent and Suzie Marcum, first-generation produce farmers.
- Learn about basic planning and concepts for start-up and growth.
- Learn simple ways to assess the market for your product.
- Learn planning tips specific to small farms.
- Uncover methods for evaluating the marketplace to understand the sales potential for your business.

## Getting Started

The facts I outlined in chapter 1 certainly show that there is room for you to live and work down the country lane of your choice, or start a local foods business where you already are. They make it easy to get behind your choice of starting a small-scale farming business. But the rest of this book won't be so easy. In fact, I'm going to ask you to work a bit right now. We'll be examining the more serious topics of how to get started and how to avoid missteps.

If you're starting a hobby farm or a small foods business, your project may be relatively simple and may not require a whole lot of money up front (besides what you invest in the land). Spending thousands on business planning and market research conducted by major firms probably doesn't make sense. Yet, when you are new to something, particularly if you are making several changes at once, sound research cannot be overlooked. A reoccurring message from everyone profiled for this book will not-so-subliminally drill the message into your head: Take your time, understand the market, and know what resources are out there to help make the transition easier.

Here in chapter 2, I'll explore two major areas: basic business planning ideas and simple market assessment techniques. I'll help you take a concept forward to a viable venture, discuss the basic first steps in establishing a business, and demonstrate why every small enterprise can benefit from sound initial planning.

I'll also delve into the concept of market assessment and explain some strategies that work especially well for direct-to-consumer marketing and how to apply them without spending a lot of money. Using this simple tool to assess the market, you'll be able to determine whether you should go ahead with your hobby idea. This chapter also sets you up for success by providing a solid base on which the remainder of the book will build.

The profile of microfarmers (farmers who have large acreage but not a full-size commercial farm) Brent and Suzie Marcum of Salem Road Farms in Liberty, Indiana, will show you that planning is essential but not complicated, fancy, or even too fussy. They'll also illustrate the all-important concept of changing plans at least every year to keep your progress paced just right—not tortoise slow and not jackrabbit fast.

# Planning

I'm inclined to call this section Business Planning Lite—not because we're on an information diet, but because I'm offering you some business planning pointers, not a five-volume comprehensive planner. So many resources are available on how to create a business plan (one of which I've written myself) that I don't think it's necessary to duplicate all that information here. Instead, I'll focus on planning topics that are directly related to hobby farms, since you may not have considered these areas before.

Here are some **good resources for business plan templates:**

www.sba.gov
www.allbusiness.com
www.myownbusiness.org
www.score.org

For a sample business plan that contains many of the components discussed in this chapter, see the Sample Business Documents at the

Brent Marcum pruning eggplants in the hoop house.

back of the book. Use this plan as one example of how to develop a business plan; it may need to be expanded for your own business.

It's also fair to say that while you'll see that planning is essential to avoid pitfalls and stay organized, if you're beginning a glorified hobby, a massive formal plan probably isn't necessary. There are exceptions. For example, if you're spending a great deal of money, applying for grants, or going to the bank for funds, a formal business plan is essential. Either way, you have to think before you act. So read on to learn the basic components of a business plan.

## Business Plan Basics

When I began Aubrey's Natural Meats in 2003, I started with a 30-page business plan. It made sense. First, as a writer, being verbose was no problem for me, and planning made me feel as if I was doing something. Second, I wanted to borrow money from a local bank and knew they'd want to see projections for returns and cash flow. Yet, though I strongly advocate having a solid plan, newbie ruralistas are not likely to write up a formal document. Brent and Suzie, for example, were taking their start-up funds out of their own pocket, and they truly started as a hobby with the hope of producing food for their own consumption.

These are **musts for any business plan:**

- ꙮ Find a template you like and alter it to fit your situation.

- ꙮ Review it with professionals. If you're making a budget, share it with your accountant.

- ꙮ If you're looking to set up a corporation or a limited liability company (LLC), get legal advice.

The structure and length of your business plan will vary widely, depending on what you plan to do. But the basic components are as follows.

- **Mission statement:** This can be one sentence or one paragraph. It's a formal statement that defines who and what you want to be and your objectives or goals.

- **Company profile:** The company profile describes everything from the name and address to the legal designation (corporation, LLC, sole proprietorship, etc.). For location, it describes the surroundings and the area or setting of the business. It lists who owns what percentage of the company and the roles and responsibilities of the participants. The profile also provides more detail on what the company does.

- **Product description:** In this section, which can be as brief or as detailed as necessary, you define and describe your product. Strong plans also detail quantities expected to be produced and time lines for production, as well as product pricing.

- **Resource/supply assessment:** This section includes a discussion of how the raw materials will be acquired, how steady the supply will be, and whether the supply will have any seasonal interruptions.

- **Market assessment:** A market assessment is a description of the desired or targeted customer and the marketplace for the product. This can include demographic information, such as profiles of people and regions where you will be selling.

- **Start-up costs:** This section should include accurate prices for everything from land to equipment and raw materials, permits to legal and accounting fees.

- **Budget:** Put together a budget for the first year to demonstrate how your start-up costs will balance out against how you plan to price your product and against projected first-year sales. (You don't have to recoup start-up costs in the first year.)

- **Growth prospects:** Anticipate your growth for the first year, as well as your overall goals for the first through fifth years.

## Meet the Crew at Salem Road Farms

"Dad wasn't a farmer; he was in the timber industry and invested in farmland," explains Suzie Marcum. "I didn't really grow up on a farm at all, and Dad wasn't involved day to day, just as an overseer."

Husband Brent Marcum, like Suzie, grew up in rural Liberty, Indiana, but wasn't raised a farmer either. He does have fond memories of time spent with his grandparents, who outlasted the Great Depression, that fostered his love of the outdoors. "My parents worked all the time when I was a kid, but I was at my grandparents' place a lot and they had huge gardens that I spent lots of time in," Brent says. "I can remember going to the orchards in the fall on Sunday afternoons with my grandparents. I also have one grandmother who is half Cherokee; I think that helped with that connection to the land."

Brent and Suzie arrived at their piece of the earth in a rather roundabout way. Brent spent three years at Purdue University studying engineering and worked a variety of jobs, from painter to purchasing agent for a bank, before realizing in his late 20s that he didn't like being away from home all the time.

Suzie, meanwhile, had a baby and moved away to attend college at Montana State University. She was a single parent at the time. "I knew with a degree in chemical engineering I would have to travel and change jobs every two or three years, so I transferred into horticulture," she says. "I also began living a more natural lifestyle and I was introduced to homeschooling." She ended up returning to Liberty, where she got reacquainted with Brent, whom she'd known since high school. The two have now been married more than ten years.

What is now Salem Road Farms sits on seven and a half acres in a U-shape that Brent calls a glacial sandbar because there are streams on two sides and forests on three sides. Suzie's father gave them the land as a wedding gift, but they didn't move there right away. "We actually camped out here," laughs Brent. "This was an alfalfa field surrounded by forest."

"We've been fortunate in that we had some savings and could pay start-up costs ourselves," admits Suzie. "We've not had to borrow money." But, says Brent, "we're still not throwing money at this. It has to pay for itself."

Their early foray into direct sales started small. "We began with bedding plants and sold them at a local hardware store in town that let us consign them, then we started going to farmers markets with them," Suzie says. "Then we started adding a few vegetables and herbs. That first day at a farmers market, we made $100. I was so excited that we could pull a hundred bucks from our yard!"

The crops and their diversity have grown just like kids and weeds around Salem Road Farms. Brent and Suzie now have more than 100 fruit trees, five or six varieties of berries, and more varieties of vegetables than one could eat in a summer. When he's selecting varietals and plants to grow each year, Brent says, "What I'm trying to do is achieve something that makes money and provides us with food.

"We can't legally call ourselves organic, because I don't pay the government [to become USDA Certified Organic], but we're raising good food without artificial herbicides, pesticides, or fertilizers," Brent says proudly.

Brent and Suzie make an effort to grow unique or harder-to-find items, such as Asian herbs and vegetables, and also to respect tradition with many kinds of heirloom tomatoes and other "antique" plantings. Large hoop houses (buildings constructed to act as greenhouses that are not usually heated using fuel or electricity, so the greenhouse is really green and solar or heat warmed) extend both ends of the growing season and help tremendously to reduce insect predation on the crops. When we spoke in late October, after the area's first frost nipped anything growing in the open, Brent told me, "We've still got peppers, eggplants, just about everything in there and it's just fine. I close the ends at night to trap in the heat."

They've lived on their farm for seven years. Brent and a friend built the house that he and Suzie designed. They have six children, Brandon, Kimberly, Brianna, Logan, Levi, and Elijah. Grown son Brandon has a son, Brian, and lives nearby and raises a few cattle. The three youngest boys, who are still at home, work alongside Brent and Suzie, contributing in a meaningful way to chores around the farm.

"We're a sole proprietorship and I do the accounting," says Suzie. "We haven't really used any outside professionals. We pay taxes on our profits—and we've always been able to make a profit." The profits certainly weren't large at first, and they're still not enough to sustain the family, but profits have come because their investments in the business have been small. "It's really about being able to feed ourselves and the rest, that's extra," says Suzie confidently.

## Plan for Growing Seasons

How long does it take for raspberry bushes and grapevines to produce fruit? How long do goats need to roam before they have kids? How long should cheese age? This is where planning for growing seasons comes in.

Brent advises starting your homework on the segment of small farming you're interested in long before you dig the first hole and order the first plants. He's absolutely right. Everything from the region you live in to the breed of plants or animals you buy affects its ripeness or maturity. It's essential to understand the seasons of growth as they relate to your new venture.

When Brent and Suzie started buying fruit trees at $25 apiece, they already knew what they were getting themselves into. "With our fruit trees, we had to plan ahead three to five years for a crop, so we knew the return on them would be later," Brent says. Knowing that, they didn't overextend themselves.

Take courses, read reliable information online, and talk to other people who've gotten started recently or have years of experience. "Put the seed in the ground and it will grow" is not going to work.

"You just have to realize that there are some things that are going to go great and others that you might as well forget after you've tried them," says Brent. It's **good business advice**—and good advice for a lot of other things, too.

## Plan for the Changing of Seasons

I think you'll be amazed when you move to the country at just how much more you notice, feel, and even *sense* the seasons changing. You can smell the earth so strongly in the spring, and the hint of rain practically permeates the house. On July days in the Midwest, you can hear the corn stretch and grow in the sweltering, humid air. By fall, the sight of combine headlamps amid the dust and red light of dusk will become a sight you know means harvest. In winter you can feel in your bones a cold that signals a snow squall, and the gray-blue late afternoon light just begs you to finish up your chores early and go inside. Yes, the seasons affect us deeply when we're *out in them* rather than just driving through them.

When you're starting a hobby farm, be sure you're ready for the changing seasons. Resources like feed and bedding are often purchased early for animals, and in tough years can be hard to locate after the first snow falls.

Crops and produce must be gathered before the first frost, and it's hard to know what that date will be, even if you've bought a Farmer's Almanac. Do some research about when each season typically begins in your region and what you'll need at the change of seasons. Make sure that whatever work you do for your main income is compatible with the seasonal nature of your farming life.

## Planning for Seasonal Cash Flow

Just as the crops and animals and foods such as meat, wine, and cheese are seasonal, so, too, is your cash flow. Cash flow is so important when you're starting a farm-based business. It's not like your nanny goats will be handing you a regular paycheck every two weeks—but they'll still expect to be fed every day!

Where you sell your goods may also be seasonal. Evaluate your selling venues (I discuss this at length in chapter 8) and know ahead of time when these places open and close. Then make sure your expenses match up with your markets.

In the winter, the dwarf fruit orchard at Salem Road Farms is pretty but not productive.

## How Brent Plans

Each year, Brent plans his farm anew. "I kind of walk around, looking at what space I've got and looking at those items I think might sell well in the future," he begins. Allowing yourself time to take a holistic approach to the planning process really makes sense when you're on the farm. You're not planning just for your business, but for your lifestyle. Realize that your plans may also affect the land and the environment.

At Salem Road Farms, Brent looks closely at his land and what is possible on each part of it. "Our property has both northern- and southern-facing slopes. Therefore, I take into consideration the microclimates of each. Those southern slopes can add an entire climactic zone for the right crops."

He continues, "We expand gradually each year because otherwise it would be overwhelming." Suzie adds, "There was a steep learning curve and there have been failures, mostly due to neglect, because of expansion too fast." These were largely due to planting too much, in which case weeds or insects took over what they couldn't keep up with.

"The first three years, we had certain areas around here that were disasters," Brent admits. "Now we're planning for more permanent things," such as fruit trees and bushes. "These things carry a higher investment, plus the time it takes until they produce, but they don't take as much work."

Brent and Suzie started conservatively and inexpensively, and have no regrets about that. They recommend buying used equipment at first and buying only what you know you'll need right away. They started with a small hoop house for the bedding plants that Brent put up himself; the building cost them only $200. Over the next few years they added more buildings, each a little larger, but always with the idea that the addition would pay for itself in the first year after purchase.

"The big thing for start-up is extensive reading, I think," Brent says of his learning curve. While he didn't attend any seminars, the list of books and authors he's read would fill an entire book index.

What will you do to cover expenses when your seasonal farmers market closes? Do you have a plan for saving or a place to deliver products in the winter? Will you shut down production totally or save cash, or should you be doing something to create inventory for next spring? Much direct marketing of food items is based around the *farmer's* schedule, not the consumer's. Consider that you're now on that time frame, and make sure you have a way to stay in business.

## Transportation and Distribution

Most business plans include a section on how the product will be transported and distributed. But products produced on hobby farms require special consideration because your channels for getting the product out there are more limited due to your small scale.

In short, "more limited" means more labor for you. Make sure you understand direct-to-consumer marketing and are familiar with the outlets in your region for your product. Evaluate the time it will take to get from one location to the next and the frequency with which you'll have to replenish products before you make agreements with too many stores, restaurants, or consignment shops.

Look into getting a **retail merchant's license** in your state. When you have one, you may be able to qualify for tax-exempt diesel fuel. It's worth checking out.

## Labor

In the summer of 2008, my husband and I hired our first intern to help us run the meat business. We were nervous about bringing someone in and opening our home and business, but the results were fantastic.

Rural life requires more labor than you can imagine. Seasonality and the weather can cause everything thing from droughts to a never-ending crop of weeds. When planning for your new enterprise, seriously evaluate your labor needs and decide if you can commit to the amount of time the business and its management, marketing, selling, and maintenance will take. If you're not sure, seek out other entrepreneurs who are doing the same thing and ask them to provide a sketch of their day-to-day workload and how much time each task ordinarily takes.

# *Evaluating the Market*

Like the planning section, this section is an experience-based discussion of market research that I've created specifically for hobby farmers and new foodies. It is not a detailed tutorial on how to do market research. Yet I can't underscore how important knowing your marketplace is to the success of your new opportunity.

The Marcum boys, Levi, Elijah, and Logan, help out with the seedlings.

You're coming at this market from a consumer's perspective, and that is a major advantage, especially if you're starting a business to respond directly to a need you feel is not being met. Understanding (and caring about) consumer preferences is an area where many lifelong farmers fall short. You don't have this problem, so on some level you've got a built-in sense of the market. Still, when it comes to knowing what will and won't make a go, you've got to learn the market from the purveyor's point of view, being conscious of supply, market saturation, price points, demographics, and consumer knowledge and acceptance of your product.

## Peruse the Market

Go see where you want to be. This rule is simple, but surprisingly, many people overlook it. I can't imagine selling in a new farmers market I hadn't visited before, but many vendors do it. I think they're making a mistake.

Many land grant universities (those with departments of agriculture) offer short brochures and publications with **advice on how to start up an agricultural business.** Though not comprehensive, these quick reads may give you a few tips you hadn't thought of before. Try the publications offered by Purdue University's New Ventures Team, found at www.ces.purdue.edu/new/.

Get to know the markets where you plan to sell by visiting them more than once.

Spend time learning about the marketplace where you plan to sell. Visit stores, boutiques, markets, restaurants, or any venue you're considering. While you're there, don't just look and listen; talk to people. Talk to other vendors, talk to customers, meet the managers, and get a general feel for the place.

There are long lists of formal market research techniques that you can use to evaluate a particular market quantitatively or qualitatively. I've written about these techniques in other places. You can build a questionnaire to e-mail to prospective customers, friends, business acquaintances, and others whose opinions you respect. You can conduct focus groups with peers, potential customers, or potential retailers. You can interview influential people in your community. (Look in the Resources section for more places to discover ideas about this technique.)

Even if you're conducting your research informally, you need to be sure to ask certain questions—of others and yourself.

- How long has the venue been in business?
- Do you like the layout and design?
- Is there adequate parking?
- Do customers appear happy?

- What's the ratio of repeat customers to new "walk-up" shoppers?
- Are people actually shopping or just browsing?

When you're checking out a particular market, be sure to visit several times. If you're planning to sell bedding plants, the way Brent and Suzie started out, that's an early spring buy. Don't set yourself up at a farmers market nobody goes to until it's sweet corn and tomato season. And how will you know this if you don't visit the market more than once?

Statistics also help. Visit local community development offices or the chamber of commerce to find out about traffic patterns and size for the roads you'll be located near. You may also find out population data and demographics such as mean income that can tell you whether the local population has the means to buy what you want to sell.

## How Brent and Suzie Study Their Market

Like most of their peers who market food directly, Brent and Suzie use a variety of techniques and avenues to distribute their products. As they've learned the marketplace and continued to conduct their own market research, plans have evolved and locations changed. Brent is gone from time to time on interior painting jobs, and with Suzie homeschooling three children and keeping up with the crops, their markets have to be well-targeted and efficient.

They stick with the best spots, such as their local farmers market in Oxford, Ohio, and let some things go, such as a farmers market that was just too long a drive for the money they made. Brent and Suzie have also evaluated farmers markets based on seasonality, finding one that went only from May to August too short to fit their extended growing season.

"Occasionally we do sell to a couple of restaurants, and we know a family who goes to a market in Cincinnati that can take our extra stuff for us and we do a share deal with them," Brent says.

They've had two hardware stores selling their bedding plants at once, but cut back because handling and delivery time got to be too much of a drain on Suzie's day.

Brent and Suzie spend time looking at customer buying habits in terms of trends that are individual to their products but also trends for their local area. "We don't do any formal research, but we do keep track and experiment with products to see how they sell," Brent says.

## Do Your Own Research

In addition to visiting sites and asking questions, you can and should do some research on your own. There are so many online resources that it would be foolish of me to try to list them all, and the convenience of the Internet is unparalleled. Visit sites that offer the same product you'd like to produce and sites for other vendors in your area. Look for which products are overloaded and which are missing. Fill the gaps and you'll find business.

Reading this book is another form of market research, by the way, so keep reading and, when you're done, look for more books. Brent has been studying his market for more than ten years and has no intention of slowing down just because he's learned a thing or two. Read.

---

They've sold through a cooperative store at a college campus and sold eggs for a while, but found eggs and chickens lacked the profit margin to keep them interested, even though customers liked the product. They've also conducted tests on selling some produce items at the local grocery store and found that customers really liked the convenience.

Brent and Suzie have also taken a look at the demographics of their customers. The Oxford farmers market is one example. "There are 330,000 people in the county and a university in the town's population base," Brent says. "They want fresh and local, and we're more successful there."

They find that customers are interested in convenience and often need food prep ideas, so they offer several recipes along with the produce needed to make a dish. If they don't have all the ingredients, they're willing to suggest another marketer who has the item.

Though they have many upscale customers willing to pay premium prices, some consumers are price conscious. In the past they'd eaten all the produce that "wasn't that pretty," as Brent calls it, but now they offer products that are less than beautiful at half price. Selling with catchy titles has help moved the less attractive produce; they've used signs like "Tomatoes with Troubles," "Peppers with Issues," and "Pears with Problems." It works!

## Why Do People Buy from
## Salem Road Farms?

"There is a real trend of people wanting to have their own garden, or feeling like they should have one. But so many times they fail because people don't realize what an investment in time it is," Suzie says.

Early in the season, plenty of customers tell her about their big ideas—only to abandon them by the time the hot weather arrives. This inability to tend their own gardens is just fine with Suzie, because it means more business for her. It also seems to bring her customers closer. "If you know your farmer personally, I think for some it's like growing your own garden," she says. "It kind of takes the guilt away for not producing their own food."

Networking is also a great research tool. Networking is more than shaking hands and eating a mediocre lunch at a conference center; it's about building a base of contacts you can turn to again and again. It's about getting answers to questions you didn't even know you had when peers raise

My husband, Cary, spends time in the office planning ahead for our business.

## Remember What's Important

"If you focus too much on the business, you will lose sight of your lifestyle," Suzie says. She and Brent appreciate their second income and find the money really helps, but they guard against rapid growth—anything that takes the focus away from having a happy family and contributes a feeling of being too "corporate."

Several years ago Brent had an epiphany when he went to Mexico. He noticed that around Cancun's hotel district people lived a highly materialistic lifestyle but seemed stressed. As he got out into the country, he saw tiny homes and subsistence-level living, but the people seemed happy. "I realized when I got home that we don't really need all this stuff to survive," he says plainly.

Brent and Suzie know that growing their business takes work, and for them, it's not always work they're willing to do. After all, Salem Road Farms was started to be a lifestyle first and a side business second. When I asked Suzie if she had a web site, she said no. "I don't want to market too much," she explains. "The web might help grow us too quickly!"

pertinent issues. And it's about staying motivated when you begin to get weary. Search for local trade groups and attend at least one industry meeting before you start your farm business.

Taking courses can also help, especially if you're going to raise or plant something you've not only never seen before but certainly never cared for as a living thing. Calves are cute, but did you know that it can take as long as two years for a grass-fed cow to be ready to become beef? Get educated—in advance as much as possible.

## Just Get Started

Your business planning and market research may seem mundane, or even something you'd rather not bother with at all. But this is important work. Let's relate it to agriculture and the seasons.

Planning, and then waiting for those plans to unfold, is like your own springtime. Just as the grass waits to turn green in the spring, you may be dreaming of blooming flowers and other lovely warm-weather transitions as you lay the groundwork for your business. Thinking of your new venture critically yet creatively will enable you to uncover the fresh grass of prosperity and potential underneath the blanket of snow. So I leave you with the idea that if you've worked hard to plan and taken the time to assess your market, you're almost ready to bloom.

Business planning is an exciting time. I spent a lot of hours planning because I found the process so encouraging. Just learning about the prospects and thinking through the potential challenges that lay ahead stimulated my mind and kept me motivated to prepare to leave behind my career in town. The zip of electricity you feel when you start to set your plan in motion is like a first kiss. It's so gratifying to be *in the process* rather than just mulling an idea around in your mind. I encourage you to start planning and investigating, and start the journey.

I'd take the journey again if I could. But it's your turn now.

## Brent and Suzie's Best Practices

- Research and read as much as possible before you start.
- Don't quit your day job too soon; keep cash flow estimates conservative.
- Consider the large amount of manual labor required on a small farm.
- Have a liability insurance policy for your small business that is designed specifically for your industry.
- Be sanitary in picking and transporting the foods you sell. That includes not using recycled or reused goods just to "be green"—safety first!

# Chapter 3

# Funding Assistance

**Learning Objectives**

- Understand the different funding options available, including their pros and cons.
- Learn about grants, rural development programs, and other kinds of financing resources.
- Get tips on how to apply for grants.
- Learn about cost-sharing programs available in most states to help with marketing expenses.
- Learn about cost-sharing programs for maintenance and farm improvements.
- Read about cattlemen Jon Bednarski and Dan Weintraub, who use grant and funding programs to help their business thrive.

## *Be Creative*

From personal experience, I can tell you that how you fund, or fail to fund, your hobby farm or local foods business is just about the most essential step you can take when embarking on this journey. Yes, in chapter 2 I espouse the benefits and necessity of knowing your markets and creating a solid, realistic business plan. But I'll be honest: Plans can and should change.

Burning up too much cash will splatter the best-laid plans like an egg dropped on hot pavement. It's not going to be pretty.

You need a way to access cash or bail yourself out, if it comes to that. I'm not suggesting that you spend years just dreaming about your business, stockpiling every dollar you might ever need to be successful. *Au contraire.* As a "serial entrepreneur" (something a dear friend once called me), I don't have the patience to wait for everything to be perfect before proceeding. No, I'm not suggesting that at all. Rather, I believe that funding your new adventure *creatively* is the way to play—and keep playing through the tough times and the great times.

As I write, we are in the middle of what has come to be known as a credit crunch. What if you're out there trying to make a lifestyle change and your stock portfolio is taking a nauseating roller-coaster ride? Or your home's value has plummeted, or interest rates are so up and down that just opening the business section of your newspaper makes you dizzy? I don't know what the markets will look like when you pick up this book, but I know the question that's probably rattling around in your mind because so many people are asking it: Is there any money left?

Markets will ride wild curves. But despite the worry these curves can cause, I'm guessing you still want to make a change. Does your dream have to wait until the uncertain economy has calmed down?

I was a financial planner for a midsize brokerage firm before I gladly left in 2004. The skills and knowledge I gained there served me well when I started my farm business, and people still ask me a lot of financial questions today. One of the main things I learned was to plan carefully. So I certainly can't make a blanket statement that anyone concerned about investments or assets should throw caution to the wind. Nope, you're not going to hear me yell, "Yes! Leverage it *all* for your dreams as you go for broke!" In fact, when you get to chapter 11, you're going to hear me encourage you to use caution and start slowly enough that in your first year you can hope for *failure avoided* rather than *success imminent.*

Here's the thing, though: Money is available no matter how the economy ebbs and flows. Finding it and being creative about getting it is how you gain access to the funds you need.

There are a lot of resources available that explain how to finance a start-up business. So in this chapter I'm mostly going to present what I hope are new or at least underexplored ideas. I'll help you understand financing options for small hobby farms and local foods endeavors. Because traditional bank loans may not be as readily available for these types of projects, my emphasis is on applying for grants and offering shares in your business.

Wintertime brings challenges and delights at the farm.

## My Own Funding Story

I started my small farm business in late 2003 with all the vigor and deter-mination of a zealot preparing for battle. I looked at data and competing companies. I studied the market and papered it with questionnaires. I went to conferences for two years before I opened, and I used reams of paper and years of intellectual capacity drafting well-crafted business plans. I made some fancy financial statements that looked ultraprofessional. I was astute and diligent. Then, knowing that I needed some money, I went to the local bank.

That's it. I just called them up, told them what I wanted to do, and began (with my typical zeal) busying myself with forms and loan applica-tions and number tweaking until I was certain I'd get approved for a line of credit. And I did. No big deal. I had an idea, and I had some cash to go.

Clearly, the story of how I financed my business is not particularly unique, didn't take me outside the box, and didn't force me to uncover new resources. I just went with the old tried and true. But that was a mistake.

It wasn't long before I was starting to sell my product while still work-ing full-time. Life was going fast, and I was stressed out. I thought that my experience as a financial planner qualified me to run a small business. Absolutely hilarious—or at least it would have been funny if it hadn't cost me so darn much.

Anyway, I never reached out to anyone that first year. I was the entrepreneur and I knew best. Yes, there were resources out there: Small Business Administration and state and federal grant programs, workshops and training given by groups from farm bureaus to women in business, but I didn't look into them. In 2004, a friend in rural development even encouraged me to apply for a value-added grant program, saying I could get up to 50 percent of my initial start-up costs paid for. I convinced myself that I didn't need grants and shelved the info in the famous File 13 (the trash can).

Because I didn't reach out to network and I never sought a cost-sharing option on financing, by the end of my first nine months in business my line of credit was tapped out and I was temporarily paralyzed—unable to make my next expansion. Other factors pushed me to this cliff, including costly errors made by an awful subcontractor who overpromised and underdelivered so badly that I spent hours doing work I paid him for and issuing credits to disappointed customers. (I'll give you some advice about how to avoid these kinds of problems in chapter 10.) Still, if I'm telling the truth here, I was the entrepreneur in charge, and I made sure that I didn't have anyone to turn to.

So guess what I did? I went right back to the bank and applied for a little more cash. This time they responded with less enthusiasm; they said no. To me, the "no" was crushing. It wasn't even so much about the money; it was that, for the first time, I realized I'd made a mistake in believing you don't need any advice or help when you start your own small business.

The letter I received denying me an increase in credit literally wiped the smile of foolish pride right off my face. I was wrong. I stood on my own, all right, but I didn't have any place to turn for help. I never even told my husband that I had applied for a second line of credit. And I was so ashamed when I didn't get it that he won't know about it until he reads this book!

You're probably asking yourself, if I knew about other sources of funds and even people who were willing to help, why on earth didn't I pick this low-hanging fruit? Well, I just don't know how to answer that. Certainly I felt at the time that leaving my "town job" to make it in the country was so empowering that I didn't need those other options. Also, I think we sometimes tend to scoff at government programs—I certainly used to—thinking they are a form of welfare for someone else.

When I first started my business, I was competitive, lacking the spirit of cooperation I've developed over the past few years. I was afraid that if I applied for a grant or cost-sharing programs, somehow I'd be offering my personal trade secrets to everyone for free. Life's funny, isn't it? Now I'm sharing my best practices with anyone who buys my books or attends my seminars. The bottom line is that I should have reached out early on for help. I chose not to because I didn't understand the benefits.

## How It Worked Out

Of course, if I'd lost everything, I wouldn't be able to write this book today. I changed my business plan, moved entirely to direct-to-consumer marketing, and put in some cash from my own pocket for a time. I also reached out, going to meetings and workshops and generally looking at what was out there.

It took only a couple of months before my business was rocking and rolling. We were bringing in cash, and I felt redeemed. My mistake in not seeking funding assistance had been expensive, but it wasn't the end of my local foods business. The frightening thing is that it could have been.

## Why Air My Dirty Laundry?

I'm sharing this experience with you not to cleanse my conscience but to convince you not to get so caught up in your desire to live the independent country lifestyle that *you* don't use the resources that are out there. This chapter is about resources and ideas for unique funding opportunities. But knowing about them won't do you any good if you don't use them.

Going it alone is not the only answer, as you'll learn from the farm I've profiled in this chapter. After reading the story of Sherwood Acres Beef owner Jon Bednarski, you'll be searching for programs the way insomniacs channel surf at 2 a.m. There's money out there. Now let's look at where it is.

# *What Funding Is Available?*

You already know from my own financing story just how important I believe it is to use creativity and seek assistance when you start a hobby farm or foodie business. This section briefly goes over some common types of funding and their pros and cons.

The basics of funding any type of business are the same across most industries, which is why I'm not going to cover them in great depth. There are hundreds of books that can help you with traditional sources of financing. I'll just remind you of what those sources are and offer my comments and experience.

Jon Bednarski, his wife, Sylvia, and their children, Kristin and Kyle, on the farm.

## Borrowing from the Bank

When it comes to traditional banks, the most basic options are:

- **Term loans,** which commonly have a fixed rate of interest and a contracted term of payment
- **Lines of credit,** which can be drawn on and repaid over and over, but often have variable rates of interest tied to the prime rate

While I relied too heavily on this source of funding, it has many benefits. When acquiring assets such as property or equipment, or making any sizeable purchase, most new venture owners take out a loan. The application process is fairly straightforward, and the repayment terms are easy to understand.

A great way to use bank financing, especially from year to year, is to combine it with another program, such as a grant, using the loan money to match the highest amount of grant money you can receive.

The suggestions and ideas I'm offering are based on my research and experience, not those of an accountant or a licensed financial planner. **Consider each funding choice carefully** and evaluate each one as it pertains to your unique situation.

Bank financing, if you use it correctly and have a good repayment record, can also be essential in building a solid credit history and a good credit score.

### Loan Guarantees

Many federal and state agencies offer loan guarantees to help borrowers secure credit. Guarantees are not the loans themselves. The guarantee is a promise from the agency to the bank to pay back the loan if the borrower defaults. Guarantees are usually not for the entire amount borrowed; typically, they cover between 50 and 80 percent of the loan.

The actual loan, along with the terms and interest rate, is typically worked out between lender and borrower. As long as the bank is using a legal rate that the borrower can realistically afford to keep up with, the agency offering the loan guarantee does not get involved.

Loan guarantees are not free. Typically, the borrower must pay an annual fee to the agency guaranteeing the loan. The fee is based on the outstanding balance.

## Using Your Credit Cards

With all the commercials on television advertising debt consolidation, credit card refinancing, and bankruptcy as an exit strategy from debt, you might think it's irresponsible to mention credit cards as a funding source. It's my opinion that irresponsible people can abuse any kind of credit, not just plastic. Ordinarily, I recommend credit cards for short-term purchases that can be repaid with inventory sold within a matter of months. If you stick to that, you'll probably be okay. So many cards offer zero percent interest as an introductory rate, and it may make sense to take advantage of no-interest financing to help keep your cash free for day-to-day use.

Be sure to look into the **Small Business Administration (SBA) Loan Guarantee Program** by asking your banker about the program or visiting www.sba.gov to see if you qualify. The program will cost you a percentage point or more in interest, and there will likely be closing costs as well as a separate application associated with it, but with this guarantee you may be able to obtain a more favorable rate or a higher amount of credit from your lender. The SBA also offers many other services, so a visit to the web site is worthwhile no matter how you fund your business.

Another responsible use of your credit card is a loyalty program, also known as a rewards program. Under these programs, you earn points toward travel or other rewards. It

## Meet Jon and Dan of Sherwood Acres Beef

What started out as a rather expensive Father's Day gift has turned into a 40-head herd of beef cattle. Since this gift has the propensity to multiply and requires daily feeding, Sherwood Acres Beef owner Jon Bednarski truly received the gift that keeps on giving.

Jon and his family live in LaGrange, Kentucky, a bedroom community outside Louisville. In June 2003, they visited a southern Indiana farm that raises Belted Galloway cattle and left owning three head. They didn't have a trailer, so the cattle were delivered a week later.

"This book hits the nail on the head for me," laughed Jon when I asked him to participate and described my readers. "I really didn't have any agriculture background. I'd never even been on a farm day to day." Growing up in Vermont (and believe me, when you call a 502 area code in Kentucky and hear an Eastern clip to an accent, you'll know it), Jon was around animals occasionally. He even showed a dairy calf at the county fair once or twice. He didn't grow up on a farm, though, and was away from agriculture through most of his formative years.

The Easterner came to Kentucky and started the Sherwood Corporation in 2001, purchasing 50 acres near Louisville. The company was originally involved in land development and preservation. Jon and his vice president, Dan Weintraub, had strong backgrounds in the log home business, constructing timber-frame houses, and in commercial real estate.

Jon's wife, Sylvia, also had no farm background. But the two hoped to use their little bit of acreage to encourage their children to get involved with animals. Their daughter now shows horses competitively, and after Jon acquired his first few cows in 2003, he hoped his children might also like to show cattle. But they just weren't into it. Jon, however, saw a different opportunity. He decided to start a business selling naturally raised beef.

---

might pay to charge your large purchases just to earn the rewards—and then, of course, pay your credit card bill in full when it's due. Especially if travel or dining out is a huge part of your new venture or your lifestyle, earning points in a loyalty program can have nice added value.

A third option that many new business owners overlook is using a credit card for overdraft protection. Many banks offer business and even personal customers a separate card that automatically deposits funds into

"Dan and I joined forces, and with our marketing backgrounds we just thought, 'How can we sell this product?'" He and Dan had no knowledge of the meat industry, but they did know people were buying natural and local beef at a premium price.

Unlike many of their colleagues who sell premium beef direct to consumers, Jon and Dan had no prior experience with being cattlemen. "We were in marketing and sales and kind of backed into the meat business!" laughs Jon. Fortunately, when he started he knew enough to know that he didn't know enough. He enrolled in several courses on raising livestock and managing a small farm's resources. He participated in the University of Kentucky's Master Cattleman's Program and a Master Grazer Program, and joined the Kentucky Cattlemen's Association.

"We sold our first steak in October 2006," says Dan. He handles the marketing and distribution for their beef products, which are sold as Sherwood Acres Beef. Jon, meanwhile, raises the animals. Having been on the farm for several years now, he admits he's still got a lot to learn. But he is finding more time to focus on growing the business. Sherwood Acres is marketed locally in Kentucky through several outlets, including some farmers markets. They also sell nationwide by phone, and they have a web site. (Coupling marketing strategies, as chapter 8 explains, is a good tactic for direct-to-consumer business.)

"Ninety percent of our gross sales are retail customers at farmers markets or direct to consumer, and we're pretty happy with that," Jon says. New ideas include starting a meat-only CSA for their local area and possibly a year-round indoor farmers market in Louisville. Being almost "in town," as Jon says (they are in metro Louisville), their goals include growth, but in a calculated way. "We've got to take a look at our volume. We know we could sell more, but we need to see just how we want to do it," he says.

your account (up to the card limit), helping you to avoid overdrafts. If you're a solid bill-payer who never allows the checkbook to go out of balance, you may think that you don't need this feature, but especially if you'll be dealing directly with the public (and most local foods businesses do just that), you'll get bad checks from time to time. It happens to everyone. I really like the overdraft protection feature; you can avoid nasty looks from bank tellers and costly surprises that are not your fault.

# What's So Special About Beltie Beef?

Sherwood Acres Beef sells meat exclusively from Belted Galloway cattle, also known as Belties. According to the Belted Galloway Society, which was formed in 1950 to create a breed registry and promote the cattle, Belted Galloways were brought to the United States from the Galloway district of Scotland in the 1950s. Jon describes them as "primarily a northern breed and very hardy, medium-framed, polled [without horns], docile, and with very good mothering ability." When he bought his first three, Jon was already familiar with the breed due to their popularity in the East.

Belties are primarily black (though red and dun colors are not unheard of) with a large white belt around their center, giving them their name and a decidedly "Oreo cookie" look. A long, puffy hair coat also gives them an appealing style that many people love. The breed is said to have more than 4,000 hairs per square inch!

"Their novelty appearance is a big selling tool," says Jon. "Using the identity difference is a great way to capitalize on families with small children, because the kids really think the cows are cute or neat."

Belties are also known for their flavorful and tender meat, which is also lower in total fat, lower in saturated fat, higher in omega-3 fatty acids, and has less cholesterol than typical store-bought beef.

Sherwood Acres Beef is sold as natural, not organic or grass-fed. "We're natural, not [USDA-certified] organic," says Jon. His cattle are hormone-free, steroid-free, and free of growth stimulants and antibiotics. "We'll never be strictly grass-fed because there's not enough pasture with 50 acres," Jon continues. So he combines free-range grazing with a supplement of corn and soybeans.

The product is sold frozen, boneless, and in vacuum-seal packages. Jon believes packaging is especially important and that vacuum sealing is key for attractive marketing to the customer. The beef is sold in individual cuts, so customers can buy just the quantities and cuts they want.

Beltie beef does need some special handling in the kitchen, and Dan likes to educate customers to encourage their best beef-eating experience. He says, "Any time we get a new customer, we give them a cooking recommendation sheet with our logo on it. And we advise them that if they overcook their lean beef, they may not like it!"

## Offering Shares or Partial Ownership

Offering shares or partial ownership of your new venture is a common source of financing that works for many situations and can be adaptable, depending on how the arrangement is structured. The simplest way to offer shares is to exchange a percentage of the business or some share of the profits with a family member or friends who chip in to help you get started.

Sherwood Acres Belted Galloway cows enjoy a Kentucky spring day.

I encourage you to write down all the details and have a lawyer look over the document. But an arrangement such as this needn't be complicated, and it works for many situations, especially if you are starting out more as a hobby than a business without a lot of initial expense. Just be sure to make good on your promises to repay family and friends.

This method can also fit situations in which vendors or subcontractors have products or services that you'll need regularly. I caution you to first get to know the skills and quality of work of anyone you plan to partner with. You don't want to be stuck with someone you later find out has a bad reputation. This is all too common when someone new moves into the neighborhood. Ask around about any person you are considering doing business with before you make a deal.

You can also make your shares in the business very structured. Formal sales of shares require set pricing for ownership percentages, a board of directors, and an annual shareholder's meeting—all of which are probably beyond the scope of the hobby businesses I'm talking about here. But you can set up a formal system to help protect your informal investors and yourself.

If selling shares and parceling out ownership of your small farm appeals to you, consider a food co-op or a community-supported agriculture operation; both are described in chapter 5.

## Using Venture Capital

Using venture capital usually means offering a percentage of your business in exchange for outside investment. This category may or may not be applicable, because hobby farms are small—generally too small to interest venture capitalists. However, if you're pooling together a group of small farms to form an organization, a buyers' group, or even a co-op, outside investors may quickly become interested. The local foods movement is here to stay, and venture capitalists don't like to be out in the cold on what's hot.

## Using Personal Assets

The last frontier in basic financing options is using personal assets. This means that you're your own Daddy Warbucks, bankrolling the venture out of your own pocket. I'm here to tell you that, almost without question, this is likely to happen to you at one time or another—whether it's to cover a short-term cash shortage or something more serious.

Depending on how you've structured your business, you'll want to use caution taking money in and out of "the company." If you're a sole proprietorship, this is usually not an issue. But if you're a corporation, it is a no-no. This area requires an honest visit with your accountant.

Saving up to start your own business or move to the country may have been a long-term goal for you. In this case, you're well positioned to use personal assets to fund your new project. You just can't get into the habit or rely on your outside employment or your savings to keep the business alive, because at some point, you will get hurt. I guarantee it.

Your best course of action is to know how much of your own assets you want to put into the new business and work hard to make the company stand on its own. In chapter 11, I offer some tips on how to prepare for this transition and how to avoid dipping into important savings, such as using retirement account funds that will incur a penalty for early withdrawal.

Using personal assets also means using things you own. What items do you own that you're willing to use for the business and willing to tie to it for tax purposes? Here are some things to consider:

- A vehicle, especially a truck or van that you use for deliveries
- A trailer or carts
- A four-wheeler, mule, or working horse
- Equipment and supplies
- Renovations to your home or garage to house the business or storage for it
- Office equipment and computer programs

Jon (left) and Dan (right) with their delivery van, the Beltie mobile.

## Federal Programs

Consider the federal and state programs described in this chapter as ideas and suggestions. This list won't be exhaustive and won't apply to every situation. For one thing, many programs are funded by individual states, trade groups, or associations. The purpose of this list, then, is to motivate you to contact the types of organizations I recommend here, on the federal level and in your own state, to find out what is out there.

Another reason this list is just a starting point is that many programs change. Funds may be available one year and be cut off the next. Or a program may be funded by a grant that is not renewed.

Finally, all the programs mentioned in this chapter have their own specific guidelines. Some are only for people living in rural areas and won't apply if you live in town; others may be valid in small towns, depending on the population. Some programs may be only for agricultural improvements on your hobby farm, while others may be available to help you start any foods business. Eligibility for programs varies widely, so this is a general description, to give you some ideas to investigate for your situation.

A final word: Get ideas here, but do your own research. This list is only a general guide to some programs' published missions and goals. Check the resources section for a complete list of web sites and other contact information for the following programs.

## EQIP

The Environmental Quality Incentives Program (EQIP) is administered by the USDA's Natural Resources Conservation Service (NRCS). EQIP was mandated by the federal government in 2002 to provide funds to agricultural land owners wishing to voluntarily implement various conservation and environmentally friendly production practices. The program also provides technical assistance and training to help local USDA offices educate farmers on how to implement and manage the production measures they choose to put into place.

The program is administered using contracts that detail which sustainable practices are eligible; contracts range from one to ten years. These contracts provide direct payments to the eligible land owner or offer cost-matching with the producer's own money to implement the specified conservation practices. Eligibility varies greatly depending on the region and the type of agricultural business you are engaged in. Different areas, particularly those in need of environmental repair (such as creek beds or high-traffic areas where animals have trod excessively), may receive greater preference and thus a higher cost match. Newer farmers can receive as much as a 90 percent match for certain environmental practices.

You can read about EQIP in more detail on the NRCS web site, www .nrcs.usda.gov/PROGRAMS/EQIP/.

## CRP

The Conservation Reserve Program (CRP) also provides money, technical support, and training to farmers. This program focuses specifically on water and soil management. The Farm Service Agency (FSA) administers the program through local offices in every state.

One of the major aims of CRP is to eliminate soil erosion, which can aid in preventing chemicals or sediment from contaminating the area's water supply. One of the most well-known CRP programs is the conversion of farmland designated as "highly erodible" into grass or trees that create what is called a filter strip between the water resource and the farmland still in production. This also promotes the enrichment of wildlife habitats in a given area.

Like EQIP, land owners are paid on a cost-share basis and enter into multiyear contracts.

## Federal Programs on Sherwood Acres Farm

"If there's a program out there, I've found it," Jon says. "I've used the EQIP program for fencing around the farm and to keep livestock out of ponds. I've also done feeding pads for winter feeding and put in waterers. These are great programs for startup farmers because they are ongoing."

Jon found out about federal funding programs by contacting his local conservation district and county extension agent. (These professionals are located around the country and are akin to paid state or federal government employees who provide information and education to those in agriculture. To find a local extension agent, go to www.csrees.usda.gov/Extension/ and enter your home state and area.) "They filled out paperwork and all I did was sign it. If you've never done anything like this, they'll tell you how."

"EQIP and CRP pay a percentage of the project," Jon explains. "With CRP, they paid me so much per acre to put in 500 trees. Now I get a check each year to maintain them." He serves on his local conservation district board as well.

Sherwood Acres also got some relief payments for tough years such as 2007, when rain was scarce for most of the summer and fall, affecting his grazing lands. "With the drought program, I got $750 and that paid for seed; I paid the other $750 and spent two days on the tractor reseeding my pastures."

## Wildlife and Wetlands Programs

The number of wildlife and wetlands programs is startling when you conduct even a simple Internet search. I've listed some common federal programs. Searching locally for programs in this area may also pay.

The programs listed here are all administered through the NRCS but are handled at the state level. Generally, all programs require a cost match from the producer (that's you) and provide funds that can be used for actual implementation of the practices as well as technical support. To learn more, check out the web addresses in the Resources section and search those sites for your state's contact information.

### Small Watershed Program

The Small Watershed Program can be used to protect and repair the watershed or other natural resources in a local area. The type of projects that qualify range from soil erosion control to fish or wildlife habitat expansion. Flood prevention measures and general water quality improvements may also qualify, as can wetlands development.

### Stewardship Incentive Program

The Stewardship Incentive Program is aimed at forest owners who do not engage in the industrial (that is, logging harvesting business) side of forestry. The program is used to maintain forests, promote forest health, and encourage land owners to keep their land forested. It can also be used to help plant trees in areas that the agency deems appropriate.

### Wetlands Reserve Program

The Wetlands Reserve Program is used to create or redevelop wetlands. The program works by creating a long-term or even a permanent contract where the owner makes his or her land into wetlands and removes it from agricultural production. In exchange, the owner receives payments during the contract for the actual "value" of the land from an agricultural standpoint.

In most cases, the agency pays the cost of installing the wetlands, up to 100 percent. The owner still has control over the land and full access to it during the term of the contract, although use can be restricted since it has been committed to wetlands.

Look into **federal and state tax breaks** for small businesses and farmers. Some basic areas to explore are credits or deductions for individuals who start small food ventures in economically depressed areas, or businesses that are owned by women or minorities. Many of these tax advantages require some kind of certification, so check it out before you assume that just because you're a woman, you're going to get a nice refund!

There are also tax credits and deductions for improving energy use in your business's equipment or buildings, and for buying hybrid vehicles. Of course, anything in this category is best decided in consultation with an accounting professional.

### Wildlife Habitat Incentives Program

The Wildlife Habitat Incentives Program aids in the development of habitats for fish or wildlife. This program is specifically for habitat restoration on private acreage, rather than in parks or other federal or state-owned property. The agency works with the producer to draft a wildlife habitat development plan and then shares costs with the owner to create the habitat on the owner's property.

### Fish and Wildlife Service

The Fish and Wildlife Service (FWS) has a comprehensive web site with information about its menagerie of programs and incentives. It also has a complete listing of available FWS grants. Producers and organizations can access funds for the protection of local fish and wildlife populations, as well as funds for education and training on how to best interact with local habitats to preserve them while maintaining agricultural interests.

# State Programs

I recommend that you search for funding in state branding or logo pro-grams. The story about the Kentucky Proud program that Jon and Dan are part of (see the box on page 59) is a solid example of what the best of these programs can do for food and agricultural producers. The basic premise is that a state allocates funds, often through its department of agriculture, to encourage a variety of business activities for new and existing small-farm ventures and to help create and promote an identifiable state brand. The food producer, once part of the program, is entitled to use the program's logo on their products.

These programs provide money for marketing and promotions, usually picking up between 25 and 75 percent of the cost (it may be capped at a certain dollar amount). Eligible expenses might include developing a logo, business cards, and a web site; advertising; and travel expenses to attend

## Jon and Sylvia's Country Lifestyle

"It's almost become a vacation hobby—we're finally starting to make it more than that," Jon says of his home in the country. "My wife is a schoolteacher, and she comes home and immediately jumps on the mule [a utility vehicle, not an animal] and drives out to see if we have any new calves."

Jon admits that if the Sherwood Corporation hadn't been suc-cessful in the land and home businesses, he's not sure he would have been able to afford to start Sherwood Acres Beef. And now that he's part of the agricultural landscape, he has a much greater respect for farmers. "I never realized how much a farmer has to know to be in business. What other profession do you know of where you have to be as versatile as in farming?" he asks.

trade shows. In return, you'll likely be required to send in follow-up reports or be subject to inspection by the agency. State departments of agriculture may also offer programs that assist with other areas. For example, I've seen money available from state government to help start a farmers market.

Additionally, many states receive money from the federal government to aid in special circumstances, such as the "tobacco buyout" money available to producers in many states that was given to farmers not to produce tobacco and instead begin raising other agricultural products. Other programs may be available at a certain time as a result of economic conditions that affect a large majority of the farming public. Flood relief and stipends for livestock feeding after a drought are examples.

Good places to look for these programs include your state's department of agriculture, the department of rural development or community development, and even the state department of revenue. Or, as Jon chose to do, ask professionals in your area such as those in the Cooperative Extension Service.

## Trade Group Programs

Many state food, consumer, and agricultural trade groups offer money to support new or expanding small producers. These programs focus only on the product that trade group represents. So, for example, a beef association won't give you money for a gourmet goat cheese business. The advantage is that you're not competing with all kinds of new ventures for the money available in your state that year.

I've found that some of these programs will share the costs of a specific project, according to a set percentage. Others will offer a predetermined amount that you can use at your discretion for the proposed project.

# Grants

Grants and loan guarantees in the rural development sector are generally considered to be growing. The prevailing political tendency away from paying direct farm subsidies and toward funding programs with specific objectives is driving this trend. For example, new standards for renewable energy and energy efficiency are being supported in rural areas and small towns through grant programs that help small businesses and farmers develop home-grown fuels and improve energy use. Other reasons for the trend include the recognizable value to the rural economy that specialty food and agriculture businesses offer.

## Kentucky Proud

Jon and Dan have tapped into several sources of funding for promotion and marketing. Kentucky Proud, a program administered by the Kentucky Department of Agriculture, offers matching funds to producers marketing food and agricultural products. "Kentucky Proud co-oped with us on some stuff that we've used for magazine advertising, promotions, and cooking demos," Jon explains.

He and Dan believe that using the Kentucky Proud logo is very important for marketing Sherwood Acres Beef. "It's really an endorsement. That program became a success so quickly, and using that logo was pressed into the minds of people. It's a feel-good thing, but it's also a credibility thing," Dan says.

A local county extension agent put them in contact with the Kentucky Cattlemen's Association, and, with their help, Sherwood Acres has prepared a host of attractive marketing materials and displays. "The Kentucky Cattlemen's Association will actually match up to $5,000 to do beef marketing—we jumped all over that in the summer of 2006," Dan recalls. They used the money to help develop a web site, two professional farmers market displays, and graphics used on items such as business cards, brochures, logos, and signs for in-store freezers.

"Because Kentucky was a big tobacco-producing state, I believe they got a couple of million dollars dropped into their budget in 2000 from tobacco companies," Jon adds. "The government is now paying for farmers to diversify into other things." Sherwood Acres Beef has been able to get a share of that pie.

I do some work as a grant writer, and I have seen firsthand that these dollars are sowing seeds, not getting lost in bureaucracy.

## What Is a Grant?

Simply put, grants are funds that do not have to be repaid. Just because the money is "free," though, doesn't mean that there are no strings attached. In fact, many people avoid applying for grants because they don't like the bureaucracy associated with asking someone for money. It's true, there is no such thing as something for nothing.

Grants can be funded by a variety of agencies, from federal and state governments to private groups. The amount of competition for funds varies widely, from a few people to thousands vying for slices of the same pie.

Receiving a grant almost always involves some kind of reporting process about how you are spending the money. The agency may also conduct an audit—once or as often as quarterly. Grants may also require you to demonstrate your project to the public, host grant agency leaders who'd like to see your project, or even give a presentation.

> The most **comprehensive source of federal grants** can be found at www.grants.gov, which is administered by the U.S. Department of Health and Human Services.

In most cases, grants are considered taxable income. Also, under the Freedom of Information Act, the personal or business information that you submit will probably be accessible to the general public after a certain period. Check the program's guidelines to find out more about this.

Most grant programs require matching funds. *Matching funds* are the additional dollars, over and above the grant money, that will be used to complete the project. For example, if your project costs $100,000 and the grant will pay up to 50 percent, you'll need to come up with the other $50,000. The percentage of matching funds you'll need depends on the grant you're applying for.

Sometimes, matching funds can be provided, in part, with an in-kind contribution. *In-kind contributions* are assets or intellectual capital on which you can place a value as they relate to your project. If you are an architect, for example, and you don't have to hire one to design your new barn, you could estimate what it would have cost to pay someone for those services and then consider the value of your time and expertise as an in-kind contribution. Most grants place a limit on the value of in-kind contributions—typically a maximum of 10 percent. A sample grant application provided by Sherwood Acres Beef can be found in the Sample Business Documents appendix.

## Examples of Grant Programs

As I have already mentioned, the number of programs offering grants seems to be growing. Here, I've listed several popular programs offered by the USDA. Be sure to check the Environmental Protection Agency (EPA) and the Department of Energy for other programs. And remember, programs change. Especially if the grant program says it was funded with discretionary funds, it may not be offered from year to year. (Web sites for all these programs are listed in the Resources section.)

- **Renewable Energy for America Program (REAP)** promotes energy efficiency and renewable energy for agricultural producers and rural small businesses.

## New Moooves for Sherwood Acres

What's next for the ambitious little company with the Oreo-cookie cows? "We reached forty head this year, the brood cows are regis-tered, and the other half are steers," Jon says. "I've thought about going to an all-steer herd, buying weaned steers from other Beltie producers. I'm thinking about how to get out of calving cows."

On the sales end, Sherwood Acres is now at three weekly farm-ers markets and sells in a local chain of four stores. "It's been a huge year for us; we've had supply issues," says Jon. When he started, they processed only one steer a month, but now they are up to three a month and thinking of growing again.

Direct sales to consumers and to some restaurants (yes, they sell their beef to restaurants at their retail price) are their biggest sources of income. In fact, meat is now 25 to 30 percent of the Sherwood Corporation's business. "I'll be 54 this year, but I've got an endless amount of energy for farming—I really do," Jon says.

I asked him if the market for natural beef is saturated because so many new small-farm ventures raise animals and sell meat. "The beef market is not saturated yet," he says. "People still have their reasons for buying. Every time there's an *E. coli* scare, it reinforces what we do."

- **Sustainable Agriculture Research and Education Program (SARE)** is a research-oriented grant program that provides funds to develop new or changing products and production methods and funds to help train and educate others about the results of the research.

- **Small Business Innovation Research Program (SBIR)** pays, in a two- to three-stage process, for individuals and organizations to develop innovative ways to grow small farm businesses by using new crops or production practices.

- **Value-Added Producer Grant (VAPG)** program offers money to producers of raw agricultural products to convert a raw product into a value-added or further processed product that can be sold in the marketplace.

In most cases, the federal government offers the biggest grants and has more regular programs that are funded each year through spending initia-tives such as the Farm Bill. But don't overlook your state government as a source of grant programs. Many state departments apply for federal funds

The Internet is an excellent place to research funding programs.

and then have cash to divvy up among local residents. These programs offer targeted dollars, and there are fewer people competing for them than there are for national programs. As an added bonus, the applications are usually much simpler to complete than a federal grant application.

## Tips for Applying for a Grant

Your cooperative extension service, or CES, will likely be an excellent resource for tips, seminars, webinars, and a host of publications on how to apply for grants. I recommend looking into free resources such as CES before you apply for anything, so that you understand the process and can get help with program timelines. You may also be able to get free grant-writing help, or you can hire a grant writer for a fee.

Here are some other tips for applying for grants:

- Read the program guidelines thoroughly; many grant applications are thrown out because they just don't fit the guidelines.
- Contact a person at the granting agency to discuss your project and assess its appropriateness for the program.

- Get help. Seek free resources, use templates, read sample grant applications, or hire someone.

- Start early, so you can be sure to be on time. Applications are time-consuming—some are as many as 50 pages long—so leave yourself plenty of time.

- If possible, get someone else (even an agency contact if they offer a review) to look over your application before you send it in.

- Make sure your application is professional, neat, and spell-checked.

- Get letters of support for your project from local and state agencies or trade associations of which you are a member. Include these letters with your application for extra credit.

- Make sure your project's payback period (the time it will take you to earn back the investment) is as short as possible. The budget for your project will have to seem feasible to the grant reviewer; if your project will take twenty years to pay back, it may not be a good fit for the program.

- Don't waste your time—or anyone else's. Only apply if you're turning in your best application. Believe me, low-quality applications don't get funded; there's too much competition.

### Consider Hiring a Grant Writer

Just because you are an intelligent person or even a good writer doesn't mean you're a good grant writer. If you're applying for a large amount of money, it may be worthwhile to hire a professional grant writer.

A *grant writer* is a service provider who charges for his or her time, just like other professionals such as lawyers and accountants. The fee structure varies widely. Some writers charge only if you are awarded the grant, but most require an up-front fee for their time. Writers charge either a flat fee, an hourly rate, or a percentage of the project's cost or amount applied for. The fee might also be a combination of these, such as a retainer for the work with a bonus paid if you receive the grant. Some fees are negotiable, and some aren't. (To get a sense of how the fees might work, take a look at my web site, http://prosperityagresources.wordpress.com.)

The money you pay a grant writer (as well as the fees you pay other professionals who do work for your business) **may be tax deductible.** Check with your accountant.

When hiring a professional, ask a lot of questions and make sure the pro has experience in this area. This is especially important if you are new to rural small businesses. If you're still learning, you don't want to teach your grant writer about small or hobby farms. Ask the writer for references, and be sure to call those people.

# A Combination Approach

Finding funding for your hobby farm or local foods business shouldn't be painful, but it won't feel like a day at the spa, either. New approaches to funding make sense because while traditional lending is still an option, obtaining traditional financing for the entire project can be difficult.

You'll need to take a combination approach to funding. Your own situation will guide you in finding the right mix of grants, loans, and personal assets. I encourage you to take your time with this portion of your journey, because so much is at stake. Let your creative energies lead you to the right well.

---

### Jon and Dan's Best Practices

- Understand the piece of property you're considering—especially its location.
- Have the soil tested and learn about the land's production potential.
- Seek help and learn from peers and professionals.
- Get educated about your agricultural topic or goods. Take courses or learn online.
- Hang out with a mentor for guidance and to pick up tips.
- Feel equipped; you're coming at this from a consumer perspective and know what consumers want.

# Chapter 4

# Rules, Regulations, and Legalese

**Learning Objectives**

- Be aware of various legal and regulatory concerns when starting a food or agricultural business.
- Learn about liability insurance coverage.
- Consider topics to discuss with accountants, insurance agents, and lawyers.
- Review regulations for selling foods and animal-based products to the public.
- Learn how to apply these rules, using the example of Bob and Carol Laffranchi and the Loleta Cheese Company.
- Understand the types of organizations you'll need to call and either notify, register with, or pay to get set up and stay in compliance.

## Freedom Isn't Exactly Free

Right about now you may be envisioning your new rural adventure as a modern, full-color, HD version of *Green Acres*. Let me start this chapter by bursting your bubble. While you may be able to step off your back porch *au naturale* with no neighbors to offend, business is still business, and you need to be aware of some constraints. Yes, it's true, everything does have a cost, even the good life in the country.

I won't apologize for bringing you both the pretty and the painful about building a small-farm hobby or business. I promised you the truth. So this somewhat dry chapter about legal, tax, insurance, and regulatory issues is a must. Read it; you'll be glad you did.

It doesn't matter whether you're a serious hobbyist or a small business owner, or whether you're starting an urban garden program or a rural cheese factory. When you are dealing with the public, putting your name out there with a service or your product, inviting people to your home or farm, and doing any of these things in exchange for cash, you have to follow some rules.

## You Have an Edge

One of the advantages of being a newbie farmer coming to the agriculture industry is that you may have more general experience with many of the topics covered in this chapter than longtime farmers who are branching out into direct-to-consumer sales. I don't mean to imply that my farming friends aren't businesspeople—they truly are. It's just that the commodity market's "grow it and sell it where it's always been sold" mentality drives commercial agriculture. Many farmers no longer think about end consumers, much less about how to sell directly to them on their turf.

In the seminars I give, many farmers who want to sell food right off the farm or incorporate some agritourism still want to focus the majority of their time and energy on production and collecting cash. They are often inclined to overlook the rules and regulations necessary to be a producer, wholesaler, and retailer all in one. It's not that farmers want to break the rules and offer products that are less than wholesome. Certainly not. It's just that farmers are folks who staunchly guard their freedom and the spirit of entrepreneurship, and their first instinct is always "Don't ask, don't tell." I sympathize.

But when you don't have a farming background, that culture may not be so ingrained. You may already be in the financial or legal industry. Or, having worked for someone else, you may be more accustomed to asking questions about how to proceed rather than charging ahead assuming that you'll do it the same way Dad and Granddad did. This, my friend, is an advantage, and it should make searching out and complying with whatever rules govern your chosen industry much easier and more natural.

At least investigate every category in this chapter before your first day in your new business. Now, during the start-up phase, is the time to set off right, rather than dealing later with a penalty or a fine—or worse yet, a lawsuit.

This beautiful wine country is not the only type of agriculture in California.

# Types of Business Entities

A *business entity* is a type of structure or design of a business for tax purposes. There are various ways to structure a business, and while some seem similar, the IRS rules that govern them vary. The two broadest divisions are partnerships and corporations. Both offer the advantage of transferring liability to the business, thus protecting your personal assets.

However, if just you or you and your spouse are starting a hobby farm or foodie business, you may not need to form any kind of formal business entity. By default, you are a sole proprietorship, and you can probably use the same income tax forms and report the income and loss on your business just as you would other kinds of income, such as wages or dividends. You are the owner, and you absorb all profits and losses, as well as all expenses and all liability.

The descriptions and ideas I'm offering are based on my research and experience, not those of an accountant or an attorney. Consider each business type carefully and **seek your own professional counsel.**

Keep in mind that establishing any of these businesses costs money. Partnerships are often relatively inexpensive to set up and cost relatively little to maintain. A few of the costs of maintaining your business entity include an annual report of some kind, separate tax documents and tax filings, and possibly some annual maintenance fees. A corporation is also required to have an annual meeting. You'll also have to pay an accountant and a lawyer to make sure you remain compliant.

With the costs involved, why bother? Business entity types such as corporations and partnerships provide a layer of liability protection for your personal assets. For this reason, even if you are a one-person show, your lawyer and accountant may recommend that you set up a different business structure. There may be tax advantages as well.

Bob Laffranchi in a field of pasture grass.

## Partnerships

These are common types of partnerships you can discuss with an accountant and a lawyer:

- **Limited Liability Company (LLC):** Liability is transferred away from the owners and onto the company. But an LLC has a much simpler structure than a corporation and is not bound by corporate rules, such as the required annual meeting and board of directors.
- **Limited Liability Partnership (LLP):** This creates a formal, limited liability partnership between two or more owners.

## Corporations

There are a lot of rules involved in establishing and running a corporation. The rules vary according to which state the business is incorporated in.

These are two common types of corporations you can discuss with an accountant and a lawyer:

- **C Corporation:** This structure is used to issue stock in a company. It's the kind of company whose name ends with Inc. or Ltd. Many farms are corporations, and members of the farm corporation have ownership and receive shares over the years. Income is taxed twice; both the shareholders and the corporation must pay taxes.

- **S Corporation:** This structure is similar to a C Corp., except the income is taxed only once. This type of company is not taxed as a business; rather, the owners report the income on personal tax statements.

## Accounting Basics You Need to Know

You don't need an Ivy League MBA to expertly manage your own basic business accounting. You do need to get organized and keep detailed

Staffers at Loleta work in the cheese factory to craft artisanal cheese from the milk of Bob and Carol's pasture-raised cows.

## Meet the Crew at the Loleta Cheese Company

The Loleta Cheese Company produces one of my favorite things to eat—cheese! I could eat cheese at every meal, so I knew that I wanted to include a cheese maker in this book. If this profile doesn't make you hungry enough to fly out to northern California, then I guess I didn't do it justice.

Husband and wife Bob and Carol Laffranchi have been making cheese for more than 25 years, and they show no signs of slowing down production on what has become a popular brand. In fact, Bob is always thinking about expanding. He comes from a dairy farming background, but he had no experience in cheese making or direct selling. Though he was a farm kid, he's a lot like you if you're considering making food from scratch.

Before I get too far into their profile, I should explain the name of their cheese business, Loleta. "The name does not come from the risqué novel," Bob begins with a laugh. "It is actually three Indian words, and it literally means 'pleasant place at the end of the water.' We're located at the foothills of the river basin that goes down to the ocean, which is three miles west of us. We really liked that concept."

Bob grew up near the little town of Loleta, California, on a farm where they raised beef and timber. "I liked the dairy cows," Bob says. He liked them enough to get a dairy science degree in college and then his Master of Science degree. After serving in the military, Bob returned home and in 1972 started teaching agriculture to high school students. "I loved teaching," he recalls. "I had great students, and I was able to work in an area I was passionate about."

The school's agriculture department was flourishing at the time and enrolled more than 300 students. One day a student asked Bob if he knew how to make cheese. Bob knew it was a diversion tactic, but decided to turn it into a lesson for his class. "I ended up giving him 15 bucks and told him to go buy a book!"

records. Start by setting up a system right now, and stay on task by entering bills, saving receipts, and noting deductions as soon as you begin. Don't wait until year's end to try to remember what you sold to whom and when. Just because you're working at home now doesn't mean that you let the fiscal watchdog in you out of the barn. Keep everything organized.

A financial management program such as Microsoft Money, Quicken, or QuickBooks is a necessity. It doesn't matter how small your business; if you

Eventually, parents with farms helped out by donating milk, and the cheese was produced as a class project, with a steam table in the school cafeteria used as a vat to make the cheese. It held about five gallons of milk.

Bob knew that improvements needed to be made, though. "We made some real ugly cheese!" he laughs. "It is actually much more difficult to make cheese on a stovetop than in a factory. You just don't have the testing equipment, the instruments, or the right cultures." Bob and his kids worked to improve the cheese for several years, and he began to look into the product more seriously. By 1982, Bob was ready to make cheese on his own.

At around the same time, tourism from the San Francisco Bay Area was beginning to increase, partly because better roads made it easier for visitors to make the 300-mile trip. "We knew if we could get something put together, we could draw tourists to us," Bob says. He and Carol started with a small factory and a storefront. Bob also hit the highway selling cheese out of his car to retailers. "If I sold a couple of packages, I got excited. It's all about adding one customer at a time."

Today, Bob and Carol make a wide variety of cheeses, including Jack, cheddar, queso fresco, and havarti. They offer both traditional natural cheese (meaning the cheese is not a processed, "cheese food"-type product) and certified organic cheese. They must use milk from two different dairies to meet the requirements for organic certification. At the certified organic dairy, the cows are not given hormones to increase milk production.

Bob says consumers still need some education about the taste and style of organic cheese, but he believes the effort is worth it. "We're in the quality food business. Some of the things we do [to improve quality] were just easy for us to do." Bob adds that he doesn't make choices that will risk his reputation for producing great-tasting cheese.

have to report something on your taxes (which, if you haven't noticed, is just about everything), you need it to be organized and accessible.

Get whatever computer program works for you and use it to track your expenses, debts, sales, and returns. Year-end tax time is not hectic for me; I simply print out my reports and give my accountant a rundown of what we have accomplished. It takes me an hour on the computer and an hour with him. Done. Don't make this part of your business life hellacious!

## Bob's Start-up Advice

Realizing that not everyone can start their food business with 300 free student helpers, donations from interested parents, and the opportunity to perfect your craft while you are at your day job, Bob has some advice for making a start-up successful.

First, find the right site with the best conditions. Look closely at zoning regulations, sewer hookups, and high-quality water sources. If you're producing a food for which animals provide the raw ingredients, a good location is essential. "We have these great alluvial soils," says Bob. "It just grows great pasture here." As a result, his cows are able to stay on pasture until late December most years, although the weather in northern California is not warm. "In the summer, the heat is moderate, too. Generally, it is very comfortable for the cows. We get some wonderful flavors in the cheese, and it all comes from the soil."

He is optimistic about the local foods movement and the number of people getting into the business—and consumers' interest in artisanal foods. But, he says, it takes a lot of work to go from dairy cows to selling cheese. "I am concerned about folks who are yearning to do something but haven't taken care of all the steps in between."

As far as learning the business, Bob spent a lot of time researching and wasn't afraid to visit with other cheese makers and ask their advice. "If you're a ways away from somebody, you're not really their competition, and they're usually willing to share," he says. Bob says he has looked into "a lot of other cheese factories' back doors" and decided that if they could do it, he could do it. "If you're smart enough to learn something, then you just have to learn it."

Ultimately, though, the decisions about your business need to be your own, and you should go ahead if you feel right about something, even if others caution you. "Don't be run off just because [an idea] doesn't work for somebody else; don't give up on your dream," Bob says. He says the best advice he got about taking other people's advice came from his dad, Severino Laffranchi. "Dad said, 'Take council from whoever you want, but then do whatever you damn well please!'"

Because you're adding a new business or income feature to your life, you'll want to contact your accountant in advance or begin working with one if you used to do your own taxes. You'll be itemizing deductions, and the best way to maximize that advantage is to hire an accountant. Write-offs, as we like to call them, can be great for reducing your overall tax burden. In fact, many small farm businesses get a tax refund in the early years due to deductions from itemizing expenses. But the rules are always changing, and I don't recommend that you attempt to keep up with them yourself. Pay a professional

You may be **tempted to do your own taxes** since you're "small," but consider this option carefully. You don't want to miss out on itemized deductions you qualify for, but you need to be sure you're keeping the right records in case you're audited. Unless you have an accounting background, I recommend you get help—especially if you're forming an LLC.

a couple of hundred bucks to get it right. Then enjoy your refund check— or at least, don't pay more than you ought to.

## Five Key Questions for Your Accountant

1. How will this business affect my retirement income?

2. Can I set up a retirement plan for my small business?

3. What does it mean if I am declared a hobby farm by the IRS? Can I still make itemized deductions?

4. Which business entity type is right for me? Can I change once I start?

5. How do I handle withholding taxes and other costs for employees or family members I pay?

# Legal Considerations

The same day you call an accountant to start organizing your new venture, call a lawyer. If you are planning to remain a sole proprietorship, you may not need your attorney's assistance to get started, but it's a good idea to run through the business scenario with a legal professional anyway and ask them to evaluate it for anything you need that you haven't thought of. Certainly, if you'll be setting up a corporation (or if you're not sure what business entity type makes sense for you), your attorney should help you file the necessary paperwork.

Often overlooked is the big picture of acquiring land and creating a business. Whether you have a young family or you are a retired couple moving to the country to relax, you need to consider your estate plan and your will when you add land or any business concern. When your company is small and not yet profitable, there may not be much to pass along, but this niche has a way of growing from one season to the next just like weeds in a garden—it can happen fast. Stay on top of planning ahead and planning for those who rely on your income.

## Four Key Questions for Your Attorney

1. Do I need to change my will and estate plans because of this new business or hobby?

2. Do I need legal expertise to set up a business entity?

3. Do I need legal expertise for transferring my business to another person?

4. Who should own the title to the equipment and farmland used for the business?

## Liability Insurance

I am not a licensed accountant, insurance agent, or lawyer. The information in this chapter is based on my personal small farm business development experience and discussions with professionals in these industries about situations relevant to creating a rural small business. This advice is meant to get you thinking about the right questions to ask professionals, **not to serve as a substitute for specific professional advice.**

Although setting up a corporation or an LLC can help protect your personal assets from seizure, insurance for your new business is more essential than ever. I'm assuming you already have homeowner's insurance or a farm policy for your residence, but don't overlook the potential (and in many cases, the likely) need for an additional liability policy.

Liability insurance is not designed to replace goods or property in the case of loss, damage, theft, or natural disaster. Rather, it protects you by providing cash to pay for legal fees and other costs associated with someone making a claim against you. It covers things such as bodily injury that occurs to someone visiting your farm or home or someone who works for you. It also provides for damage that you or your representatives might cause to others' property and covers you in the event of injuries resulting from your or your representatives' negligence.

Types of liability insurance include general liability, professional liability, product liability, and employment practices liability. There are others, but these four types cover most businesses. Many insurance agents recommend that you have a combination of a general policy and an umbrella provision that increases the amount of cash you can obtain from the policy in the event of a hefty claim against you.

## Why Little Old Me?

When I speak to groups of new small-farm entrepreneurs and hobbyists, there are always one or two folks who say, "I'm so small, we only have a few people out to the farm each year. Do I really need to have insurance for that? It's not like I'm a big company." I always answer that question the same way: "Yes, even you!"

The reason is simple: You're taking a risk (from an insurance or legal claim perspective) every time you put yourself in front of the public, whether that public is buying from you directly or indirectly (yes, you need protection even if you are a wholesaler and not an end retailer), whether they are buying from you on someone else's premises (such as a farmers market), and whether you are doing the transaction or a friend or employee is handling it. And you most certainly need coverage if you host people on your farm—even just a few people.

Liability insurance should cover you, your assets, and your company from start to finish. This is true even if your "product" is farm tours or U-pick orchards. If it's sold directly from a processor (such as my company's beef and pork), you should be covered during transit, storage, and selling. Check into liability insurance with a qualified agent and find out how much is at least a safe minimum in that professional's opinion.

## Five Key Questions for Your Insurance Agent

1. Am I already covered enough on my farm or homeowner's policy?
2. How much liability insurance do I need? Should I buy a separate policy?
3. How often should I update my coverage?
4. Do I need umbrella coverage?
5. Can I bundle the coverage with other insurance needs to save money?

## Marketing Loleta Cheese

"Hats off to people who are willing to do new things in food market-ing!" Bob declares. "I tell people who are new in the cheese busi-ness, 'Options, options, options.'"

He believes you can pave your own way, but should also use roads that are already paved, if they make sense. For example, "I am a semi-artisan cheese maker, but artisan didn't exist back in 1982," Bob says. Marketing a product using the word *artisan* is a big selling point for some, but it wasn't always so. "It helps if there is already some market penetration for your products. It is easier to bring a product to the market where the big boys have already blazed the trail." For example, offering both kosher and USDA Certified Organic products makes sense so that you can capitalize on both markets.

"Food has evolved," Bob continues. "Now it's more than just something you put into your mouth; now it's entertainment. I mean, look at all the food channels on TV! We need to ask ourselves, 'What is it that is motivating the customer?' We can influence the customer by bringing them what we think they'll like, but we can't completely choose that."

Evolving with the industry and responding to customer input keeps Bob's business flourishing. "If this business looks the same now as it does ten years from now, then I am in trouble," he says.

Building customers has taken time, too. Early on, Bob says the approach should be simple and steady. "Don't overwhelm yourself with the task at hand," he cautions. "Work on one long-term cus-tomer each day and have a good story to tell them."

Bob and his family strive to develop good relationships with cus-tomers that encourage repeat visits. His products aren't low priced, but he wants them to be "priced accordingly" so that he can afford to offer samples of everything. He also thinks gratitude goes a long way. "If someone has been into the store before, the first thing we do is thank them for coming back," he says. "Getting to know peo-ple gets more sales, and we do have wonderful sales out of our little

# Regulations to Investigate

Regulations that govern food production and handling and what we do to the environment are imposed upon us at the federal, state, county, and even township level. But even without these regulations, you'll find that getting store!" Many times, guests want to relate their past experience at the store or the events on the trips that brought them to his place. Simply talking with customers seems to make money, and that makes for a happy place to work and, of course, happy cows.

The little retail store on Bob and Carol's farm hosts about 35,000 visitors a year. Bob also invites visitors to tour the farm and the cheese factory and encourages them to stay a while by offering many food and wine items in his retail shop.

The location has a garden spot for relaxing that Bob hopes more and more people will use just to rest for a moment during their travels. "It gives people time to get out of their automobile. The world is going too fast. But you can stand there in that garden and for that blip of time you can slow down," he says thoughtfully.

Loleta offers free cheese tasting at the factory, and Bob is not stingy about it. "You can taste anything and have how much you want," he says. "People are not going to buy as much if they don't try it. The finest salespeople are the products I've got!"

He also has restaurant and retail customers, including major retailers such as Safeway. And during October through May only, Loleta sells cheese via the Internet. His web site gets about 95,000 hits a year, and Bob expects that area of the business to continue to grow.

Bob also thinks farmers markets make good selling venues for cheese. "If I was starting out today, I probably would have invested a lot of my time going to farmers markets," he says. "It is so easy to get lost in a retail store," but a market allows the producers to showcase themselves in a way they can't in a store setting.

Despite all the marketing advice, Bob is quick to say that this end of the business is not his "cut of cheese. I'm more of a manufacturing guy," he admits. He gives Carol credit for almost all the good marketing ideas Loleta has benefited from. "She deals with all the stores," Bob says. Plus, no matter how much things have changed, in most families the women do the food shopping. Which is why Bob concludes, "If you're selling food items, you should be listening to the ladies!"

into agriculture will awaken considerations about the natural environment around you like never before. It really offends me when people think farmers are not environmentalists. The farmers I know are more knowledgeable, from lifelong practical experience, about environmental stewardship than any loud, angry member of a radical special-interest group.

## Inspections and Standards for Cheese

Every cheese maker, big or small, must follow a long list of rules about equipment, ingredients, processing, labeling, and more. "There are rules, governed by the state of California, about what labels have to contain," Bob begins. Even small considerations, such as the type size for certain information, is regulated. "We submit any label changes to our state dairy inspector." Bob believes simple is best when starting out, especially if you'll be selling locally rather than statewide.

Various agencies conduct regular inspections, including the California milk advisory board, the FDA, and bodies that certify food as kosher and organic, and these inspections are becoming more frequent. "The FDA comes about every six months; it used to be about every five years," Bob says. The cost of inspections is usually borne by the producer. "In 1982, the General Fund in California paid for inspections," Bob says. He estimates that his inspections now cost him $800 to $1,000 per quarter—a cost he expects to rise. "These things will keep going up every time there is a major health blowup in the food chain. When somebody fails, we all fail."

Ingredients and cheese cultures are inspected, as is equipment. Problems can lead to slowdowns that Bob wants to avoid. He works hard to be responsible for quality and safety around the cheese factory, because even little things require verification by an inspector.

For any kind of hobby farm or food business, the list of regulations can seem long and complicated. Plenty of things are forbidden—or at a minimum, you'll need to contact someone at an agency for clarification before you proceed. The good news is that most environmental and food regulations don't change often (our food code in Indiana was written in the 1940s, for example), so once you understand and comply with the rules, you're not likely to have to implement major changes and upgrades.

Rules for farmers are mostly practical, I've found, and are based on science, not on someone's personal preferences. It's not like the homeowners' association days when a group of snooty neighbors decided that pickup trucks shouldn't be allowed on the street or in your driveway after 9 p.m., or that tacky yard decor should be banned. Nope, feel free to stick that hideous garden gnome on your front lawn if you like. And as for trucks, well, if we had rules in the country about trucks parked in driveways, I'd hate to think of the vigilante justice that might arise.

"If we break a seal on a pump, we have to call an inspector to come and reseal it. Sometimes we have to shut down for a day, and they come out to recalibrate a machine," he explains.

To make the inspection process easier and to keep his quality high and his factory running, Bob implements the standards set by his inspection agencies and respects their position on food safety. He says that most of the organizations he deals with are helpful and keep him informed of changes to food standards.

At Loleta, the team also implements many of their own sanitation practices. "We wear white pants and shirts, hairnets, and surgical gloves, and we wash our arms before working the factory," Bob says. The crew also uses specific sanitizers on the equipment every evening and different products in the morning before operations begin. His employees also must pass a written exam. Finally, Bob calls pasteurization his "biggest insurance policy" when it comes to avoiding food safety problems.

"I don't care if it's cheese, jam, or spinach, we all have a responsibility in the food chain. If you're going to get into food, you've got to deal with this huge responsibility," Bob says. He's serious about quality, and he knows that Loleta's reputation is on the line each time an item leaves his cheese factory or a retail store shelf.

The rules you'll play by address your impact on the land and on the food you produce, and the potential impact on the consumer. This section covers what I think are the areas every new hobby or business that sells food or farm goods should check into before getting started. But remember, there are regulations at every level of government, so be sure to find out what pertains to your business. Here are some of the agencies you should contact to find out what the regulations are in your area:

- **State board of health:** Some states may require an inspection of your facilities before allowing you to sell on the farm.

- **County boards of health:** Contact not just your own county, but each county your food ends up in through deliveries, farmers markets, and so on.

- **State board of animal health:** Contact them about processing and labeling requirements for animal-based products such as beef, pork, chicken, milk, and eggs.

- **State internal revenue department:** Contact them to obtain a retail merchant's license, to register and secure your formal business name, and to register for tax programs if you have employees and will be withholding taxes or providing benefits.

- **State department of environmental management:** This is your state's equivalent to the EPA. You'll need to contact them to receive permits for any land or agricultural changes you make to your property.

## Sanitation and Food Safety

Of course, we all believe in the vital importance of food safety. But practicing it on a commercial scale can take some getting used to compared to how you operate in your own kitchen. The board of health in your county and state are the places to start to purchase permits, arrange any required inspections, obtain a copy of the food code, and simply ask questions. Inspectors can be difficult, but if you start working with them with questions and compliance in mind, you'll have fewer headaches.

Interestingly, food safety codes for on-farm sales and sales in a retail shop or restaurant often differ. Farmers markets also receive special consideration and some concessions from the food code. Each county gets to make its own rules, so be prepared for differences that can seem illogical.

Most farmers markets and many festivals, fairs, and events require **proof of insurance for exhibitors.** Some will ask you to name them specifically on the insurance statement; many will require only that you provide them with a copy of your proof of liability insurance and that you keep a copy with you when you sell or exhibit. You can get a copy at no charge by contacting your insurance agent. Simply explain that you need a copy so you can participate as a vendor, and let the agent know if they must send it directly to the organization or to you.

Make several copies of your statement and keep them with you so that you can simply present one to the organization when you sign on to participate. Be sure to get a new statement every twelve months or every time you significantly add assets, products, or selling venues.

## Food Processing

If you'll be processing anything, from apple cider to beef, you'll have to follow a separate set of regulations. Most processing of animal products has to be done in, at a minimum, a state-inspected processing facility (although

Loleta takes sanitation and safety seriously. This worker wears a disposable apron, plastic gloves, and a head covering and washes his hands anytime he leaves the immediate cheese-making area. These are examples of the kinds of regulations that anyone in food manufacturing is expected to follow.

some states allow on-farm poultry slaughter of up to a certain number of animals per year). Many cheese makers produce their products at home. But an entirely separate room, divided by a metal door at least, must be set up for making and storing the cheese.

When you select a processor (such a facility that will slaughter, age, and package meats), *choose wisely and take your time.* Don't be in a hurry. Oh, did I make this mistake in the first year! And how I regretted it. I chose quickly and poorly, and the first processor nearly killed me with stress and caused major expenses and problems with my customers.

*Please do not select the cheapest, closest, most eager-to-please processor you can find.* Shop around, get referrals, and contact your state board of animal health to review potential processors' records, if possible. Ask to speak with customers, tour the plant, and be on hand to watch how they work before you commit. Also, don't commit to a long-term agreement. You may need to change, and you don't want to be saddled with a company you have problems with.

If I sound preachy on this topic, it's because I am. Processing problems (such as not getting products ready on time), constant mistakes on customer orders, deer season taking precedence over your regular business, sanitation concerns, inconsistent packaging, and employee negligence are just a few of the problems I've had or I've commiserated about with other small farm businesses. Select a processing and packaging firm wisely.

## Labeling and Label Claims

Two of the advantages of offering a product fresh from the farm are uniqueness and fresh appeal. Labeling your product with custom logos, stickers, and marketing claims can add to the aesthetic appeal and pique the

Beef carcasses from Aubrey's Natural Meats at our local meat processor. Each of the beef carcasses is individually inspected by state meat inspectors as well as local plant operators to ensure quality, consistency, and food safety. Tags on the carcasses indicate the animal's identification and weight.

customer's interest. I recommend designing your own labels as soon as you're willing to spend the money to print them.

Custom labels and statements made on the label about the product must be approved, in some cases, by more than one agency. For example, when I started Aubrey's Natural Meats, LLC, I wanted to include our logo on the label and a label claim that read, "Animals raised without added hormones, steroids, or antibiotics, and have spent their lives on pastures. All beef from Central Indiana farms."

This is an example of a label claim. *Claim* sounds like something that might be untrue, but in this case that's not what it means. It simply means that you are putting a statement, verified by you, on the label. In my case, our state board of animal health had to approve the wording, the font, and even the size of the label I could place on a package. They also had preferences about where the claim and my logo were located in relation to the pertinent information, such as the product name and weight.

## Packaging, Weights, and Measures

Whether you are selling a fresh product such as produce or a processed one such as meat or cheese, you are responsible for accurate weights and measures. In the case of processed products, you will likely be required to list the weights on the product labels. This labeling will be done at your processing plant, or at your own home if you've set up something like a cheese-making room. You will have to preprogram a label machine with this data.

That said, many farm-fresh processed products, such as baked goods, cheeses, and meats, are allowed to be sold by the piece rather than by weight. Selling this way is often easier for cash-based transactions, such as those at farmers markets. If a product is sold by the piece, the price and weight are not necessarily listed on the package. Rather, the customer selects a package and pays the by-piece price that you set.

Produce vendors are usually required to keep a scale on hand (just like at a grocery store) and sell products by weight. Inspectors periodically test scales for accuracy. Complying with weights and measures is serious business. Even at local farmers markets, fines are commonly enforced for vendors whose scales are inaccurate. I've seen vendors expelled from a market for continued compliance problems.

Many small farm businesses **store their for-sale food products at home,** sometimes doing so in blatant disregard of food code rules. Check with the health department for the county you live in about at-home storage of any food product, whether it's immediately perishable or has a long shelf life. Due to the dramatic increases in farmers markets, many of which are on weekends or after business hours for processing locations, many states now allow at-home storage of goods processed elsewhere—if you meet the inspection requirements. Check into it.

Contact the board of health in your state or county to find out about the packaging, weights, and measures rules that apply in your area. Your state will also have a department of weights and measures, often found within the department of agriculture, whom you'll need to contact to become certified.

## Permit Types

If you've ever done a construction project, you know that you probably have to get a building permit before you start building anything. On the

Loleta's custom cheddar cheese curds label. Note the Real California Cheese logo in the upper-left corner. Bob and Carol participate in this state-run branding program.

farm, there are additional permits to consider. And still more are needed for selling food products, either at the farm or through events and local markets.

### Temporary Vendor Permits

For most farmers markets, you'll need to acquire something called a temporary vendor permit. (The name may vary slightly from state to state.) You can get one by contacting the county health department and paying a fee.

If you sell in more than one county, you'll need a permit for each county. The price may vary depending on what products you plan to sell. For example, a permit to sell potentially hazardous foods such as meat and cheese may be more expensive than a permit to sell apples or pumpkins. The term length of the permit is usually set by the county. The rules that govern permits and the price you

**"The cows give us their heart and soul.** We make this great-tasting cheese. That's all," says Bob.

**Country of Origin Labeling (COOL)** basically means that products will be allowed to call themselves "Product of the U.S." While this type of label was used to some extent before the COOL mandate, the new law is supposed to mitigate inconsistencies in the use of the term and regulate record-keeping at the retail or consumer level.

COOL labeling applies to all red meats, chicken, and goat meat as well as peanuts, pecans, macadamia nuts, fruits, vegetables, and ginseng. It applies to both fresh and frozen foods. At this point, how COOL labeling requirements will affect small, local businesses that do not plan to export is unclear. Ask your local inspectors for more information, or read up on COOL regulations online before you get started.

must pay is set by the county also, so don't be surprised if one county charges twice as much as another that's practically across the street.

Most market managers will require you to buy your permit before your first day of selling and will often request a copy along with your proof of insurance. Temporary vendor permits are not the same as the type of permit required for restaurants and storefronts. If you already have a permit for that type of venue, you're not in compliance if you leave that location to sell your product at a farmers market; you'll still need a temporary vendor permit.

If you want to give out free samples, you may need a different permit or additional permits. For example, if you cook your product on site and offer samples to customers, an additional, probably more expensive, permit may be required, similar to the permit used by those selling prepared foods.

## Environmental Management Permits

On the farm, you'll need permits for many building projects—just as you would when doing construction in town. However, other activities may require approval, such as digging and excavating, particularly if you're doing so near water resources such as creeks and natural low-lying areas. The soil and water conservation service (often found through the state department of agriculture or the state USDA office) is a good organization to contact if you're considering rerouting a creek bed or doing other excavation work that disturbs the environment.

Livestock systems are particularly touchy when it comes to permits. If you're constructing any kind of facility in which animals will be kept, it's a good idea to call the state department of environmental management. Animal

manure is another big issue. While proponents of free-range husbandry consider the practice natural and beneficial to the environment, you still need to
think about how run-off and groundwater will be affected by manure and
other wastes, even from small groups of livestock or fowl.

## Bob and Carol's Best Practices

- Get advice from others, but eventually, follow your own
  instincts.
- Respect the land and environment around you and the ani-
  mals you work with.
- Regularly ask your customers what they want.
- Sell the customers what they want, as long as it works for you.
- Thank your customers again and again for their business.

# Part Two

# Fresh Business Ideas

# Chapter 5

# Creating a Cooperative

## What's Old Is Also New

Whenever I mentioned something that I felt sure would shock my grandma, she'd inevitably end up reminding me that she'd seen it all before. "You know, Sarah Beth," she'd say, "there really is nothing new."

It's true. So many ideas we now have about hobby farms, niche agriculture, and slow foods were just "how it's done" when my grandma was a girl. Food co-ops, for example, seem to some of us like a new idea. But forming membership groups where individuals work together and share the risks and rewards is absolutely nothing new. For millennia, people have banded together to form groups of like mind with like goals.

The dynamics of working together was the original reason most small towns were established. In the pioneer era in the United States, people formed banks by pooling their money, established their own stores and trading depots, and built schools with their shared supplies and cash.

Fresh produce at the Lost River Market and Deli.

Agriculture, perhaps more than any other industry, still forms organizations to foster individual and group security, growth, and profits. The dairy industry, for example, is dominated by milk producer cooperatives owned by individual farmers, as are many fertilizer, fuel, and other industries that sell the stuff farmers use. Why has cooperative membership always been such an attractive option? Because it works.

As I mentioned in chapter 1, the number of CSAs (community supported agriculture) and food cooperatives keeps growing every year. In this chapter, I'll use the example of a relatively new consumer food cooperative in the southwestern corner of Indiana to show how forming a cooperative can work. Grandma's right; there's nothing new in the basic idea. The newness comes in adapting old paradigms to 21st-century infrastructure, pace, and the needs of people today.

## What Is a Cooperative?

You already know that a cooperative is a group that's formed for a common goal. But what does that really mean? The easiest way I can describe a cooperative, commonly called a co-op, is that it is a group of people who come together voluntarily to achieve a common goal and share in the

The Lost River Market & Deli in Paoli, Indiana.

endeavors, profits, and costs involved in achieving that goal. Co-ops encourage a spirit of cooperation (go figure!) rather than individual autonomy. Ownership forms can vary, but most cooperatives are also businesses that share in buying and selling to give members market advantages on both the cost and profit sides.

A co-op can be structured as nonprofit or for-profit. It can be owned (or majority owned) by one individual who sells shares, or require equal membership buy-in from everyone until the membership is capped. Or membership may not be capped at all.

A food cooperative is usually a retail store that sells groceries and other products to consumers. (The products do not necessarily come from farmer members.) It serves as a kind of grocery store that is member owned. There are also food-buying clubs organized by individuals to

Lost River Market & Deli started its co-op with help from Food Co-op 500. The mission of this national organization is "To stimulate and support the development of new cooperative food stores in the U.S. contributing to an industry-wide goal to increase the number of cooperative food stores from 300 to 500 by 2015!" **Food Co-op 500 has a 74-page manual on its web site (www.foodcoop500.org) about how to start a cooperative venture.**

purchase food in large quantities at cheaper prices. These clubs are not necessarily businesses the way food co-op stores are, and can be organized by individuals without a membership fee or buy-in. Food clubs can be as informal as a local Bible study group or book club, while food co-ops are arranged as businesses.

## What Is Community Supported Agriculture?

Community supported agriculture (CSA) is a type of cooperative. The specifics of the definition may vary from group to group, but they all follow the same basics. First of all, a CSA takes place on a farm operation. The farm may be owned by one farmer, a group of farmers, or individual members in the CSA who literally buy in to the land. Commonly, the farm is legally owned by the agricultural producer who founded the CSA, but members have an ownership share, through annual membership fees, in a portion of the food produced each season.

Membership fees supply the farmers with the working capital they need to put in crops or keep up stock and maintain the farm each year. Members

"I just loved this area and the beautiful land here," says Andrew Gilleo, owner of Tater Road Farm near Hardinsburg, Indiana. Gilleo is one of several vendors employed at Lost River part time. He also sells root vegetables such as onions, shallots, and garlic, as well as eggs all winter.

# *Welcome to the Lost River Market & Deli*

Gourmet groceries and local foods seem to be the mantra these days for many upscale communities—at least in the suburbs. In small rural towns, these kinds of markets are not always successful or even sustainable. The Lost River Market & Deli in tiny Paoli, Indiana, is breaking the mold.

Lost River is a small co-op store that focuses on natural products of all kinds. "We're a consumer co-op, not a producer co-op, and we're dedicated to consumers and owned by consumers. But we have a large emphasis on local goods," says Debbie Turner, treasurer of the Lost River Cooperative board of directors and volunteer project manager when the co-op was being organized. The co-op opened in 2007 with the motto "Healthy Choices Close to Home."

The store's namesake is a river in southern Indiana, and it reinforces the idea that it's a local market. "The Lost River is pervasive throughout the county and is very unique," says Debbie. "You can only see parts of the river bed because it goes underground through the cave system and then literally bubbles up in the western part of the county."

Lost River Market was conceived in the minds of many interested volunteers who saw a need and wanted to fill it. The founding organization was the Orange County Home Grown group, which manages the successful farmers market in nearby Orleans, Indiana. "We had 125 to 150 local vendors and we saw the economic impact the farmers market had on those producers," Turner says. "We also saw the impact on the customers and when we saw that relationship, we wanted to strengthen it." They knew it would be a major undertaking to form the co-op, open the retail store, and then staff it daily and keep it operational. But they believed they could do it.

The co-op was formed remarkably quickly, in less than a year, and is a model of collaboration between farmers and volunteers in town. With the aid of groups such as the Indiana Cooperative Development Center, the national organization Food Co-op 500, and nearby Bloomingfoods, a successful and regionally well-known food co-op in the college town of Bloomington (home of Indiana University), the Lost River group found the professional guidance it needed.

Anyone can join the co-op, but people do not have to be members to shop in the store. Still, many people choose to pay the one-time membership fee. "About 60 percent of our business is from members," according to Brad Alstrom, the store's manager. "It's just amazing to watch people commit; a lot of them understand that if they sign up, it makes this happen." Members get a discount at the store, but they also have to commit to supporting the cooperative with their food shopping dollars. That means making the choice to regularly shop locally and support their food co-op so it will stay in business and thrive.

Vendors can also be employees at Lost River Market & Deli. One example is Espri Bender-Beauregard of Brambleberry Farm, three miles east of Paoli. She and her husband, Darren, sell eggs and lemongrass in the store during the winter. "I also make hardwood kitchen utensils on the rainy days," she says. When they work at the store, they receive an hourly wage. Espri works in several areas, including on the co-op web site and on general day-to-day administrative tasks.

"It's been a difficult summer of 2008 with the changes in the economy, but we've seen that we compete more on the type of product we sell here, not so much on price," Brad says. Their major competition is Walmart and Whole Foods Market across the river in Louisville, Kentucky, but Lost River offers a better variety of goods.

Brad himself has been surprised by the types of customers who shop at Lost River and attributes their interest in part to the commitment of being members. "Our customers definitely don't all meet the standard demographic of natural foods buyers," he laughs.

Brad believes the increased acceptance of natural and organic goods has helped fuel growth in Lost River's membership and sales. Bringing new types of products to the area that weren't available in other stores has interested shoppers.

Brad and the store's other employees and organizers believe that being a part owner in a community-building project is one of the most important factors to many of Lost River's shoppers. "It's about community. We sell ourselves as community owned; people are looking for alternatives to large box stores," he says. "This market impacts a lot of people. There is a sense of community worth and the knowledge that they're helping their own community by shopping here."

Growth is important, and Brad is noticing a synergy emerging in his local area. "It's driven by both directions; on the consumer side and with the grower who wants a market. We like to think of ourselves as really connecting those two."

share the risks with the farmer, such as crop yield loss and weather problems that could affect the amount of food they get to take home.

## Do the Differences Matter?

It depends. Since both food co-ops and CSAs are about creating a membership and distributing shares, there are more similarities than differences. However, I'm inclined to say that the difference between a food co-op and a CSA does matter, because the two differ in structure. You should know about both, anyway, to understand the significance of the co-op movement and see the real impact these groups can have on a local economy.

To sum up, in many cases a CSA is owned by the producer, while the shareholders own a portion of the raw agricultural crops. The only people who receive the bounty are paid members. A food co-op is always owned and organized by consumers to sell food and other groceries or supplies. A food co-op is wholly owned by the members who buy shares, but nonmembers can shop there—although without the benefits of membership.

## What a Co-op Is Not

With all these definitions (and their inherent similarities), it sounds as if almost *any* organization qualifies as a co-op. Untrue. As explained in chapter 4, specific rules govern the establishment of business entities. Other arrangements where people come together to conduct business, such as partnerships, limited liability companies, and corporations, are not co-ops.

**Comparing CSAs and food co-ops,** Brad says, "There are more similarities than there are differences. CSAs have a growing season (at least in the Midwest) and we operate all year. CSAs could be seen as a sort of buying club, but as a store, we're providing the convenience of a retail outlet. We have the overhead. And we provide not just local products or just food, but a whole grocery basket of items."

A CSA and a food co-op could be set up near each other or as part of an entire community plan. "Many consumers and growers have done both," Brad says. "In that sense, the two go hand in hand."

# The Basics of Starting a Cooperative

CSAs and food cooperatives differ slightly in the legal requirements for shareholders (in the case of a CSA) or member owners (in the case of a food co-op). But the initial planning can be similar. So are the steps to getting started. Keep in mind that for tax purposes, you need to hire a tax professional when forming any type of cooperative.

Here's an outline of the basic process:

- **Generate the idea:** A co-op is formed in the minds of a few innovative and interested people. Many CSAs are formed by small farmers with different products who want to create a convenient outlet for their goods, or by one farmer who needs a way to get people to take product regularly while finding seed money to keep his or her business running well.

- **Meet and discuss:** The simplest way to begin is to meet. The initial meetings need to address basic questions: What is the need, and how will forming a co-op in our local area meet it? Select a steering committee and outline who will do what in the planning process.

- **Conduct a feasibility study:** A feasibility study is essential to starting a cooperative. The study should address questions about the viability of the marketplace, potential member interest, location, and financial planning. A grant could aid with the feasibility study. Check cooperative development programs in your state.

- **Incorporate:** If the results of the feasibility study are favorable, it's time to incorporate and set up the structure and governance for your membership. If you haven't done so already, create a board of directors and, in the case of a food co-op, discuss staff positions and retail management.

- **Form the business and finance plans:** The financial and business planning stage should come right after the feasibility study. This study should address the amount of money needed and determine sources for funding.

- **Recruit members and other vendors:** Member recruitment is important early on; it can start even before the planning process is done. If you're creating a CSA, consider talking with other farmers who offer products that you do not, to increase shareholder interest and options. For example, if you have produce and beef, consider finding vendors for honey, locally roasted coffee, herbs, or baked goods.

- **Set funding goals and get the money:** Undertaking a successful funding campaign is what gets the wheels turning. This phase should include establishing reachable funding goals and giving board members responsibilities toward helping to reach those goals.

- **Prepare the location for opening and liability:** Preparing the location includes remodeling facilities and sprucing up the farm, but it also necessitates a serious discussion with an insurance agent about liability planning. If you live in the country, talk to your neighbors about the traffic potential on CSA pickup days, and ask for their support.

## Membership Structure

The cost for members and shareholders varies from co-op to co-op, and there are no rules that determine what the price should be. Your local situation, the interest of local residents, and the affluence of your area will all factor into the fee structure. In general, food co-ops charge a one-time

## Lost River's Management Structure

While producer involvement is important, Lost River Market & Deli is a retail store, and, like most food co-ops, it is operated as a business complete with a general manager, Brad Alstrom. Other employees do all the jobs that would be done at a regular grocery store, such as marketing, accounting, inventory management, and, of course, ringing up customers and stocking shelves. The store also receives day-to-day help from volunteers and even some of the local growers.

Brad's journey to Lost River has taken him almost coast to coast. He's originally from Detroit, and he had no agricultural background when he moved to Oregon and became a food co-op shopper. Eventually, he started working at the local food co-op and found the structure very meaningful. He started banking at a credit union and joined a child-care co-op for his young son.

Food has also played a role in his journey to the Lost River Market. "I've always had a love of cooking and learning about nutrition," Brad says.

Under strong leadership, Lost River is doing well. It passed its goal of having 500 members in the first year, and membership currently stands at 582. A membership costs $90, a one-time fee.

The co-op has even started its own brand, Eco-Grow. "We've developed our own little label at the local level," Brad says modestly. "It really just says that the local grower's product is raised without pesticides."

A Lost River Market employee talks with a co-op member.

membership fee of anywhere from $50 to $250. Some co-ops charge an annual fee, and others allow new members to pay the one-time fee in installments.

For a CSA, the amount of produce and the variety of products from the farm and other vendors who offer goods help set the price, as does the length of the season. Membership in a CSA usually costs more than $100 a year per household, and may cost as much as $500.

Many CSAs get the entire amount up front in the winter, before the growing season, so that they can pay for what they put into a crop. Others collect money over time to help members budget the cost. Some CSAs also offer a work-share option, where a member receives a discount or a single-season share at no cost in exchange for a certain number of hours of farm labor.

CSAs usually adjust their share price from year to year, while food co-ops often set a lifetime membership fee that is paid when a person joins.

## Pitfalls to Avoid

Co-ops aren't perfect, and the "share and share alike" feel that they inspire doesn't translate into bliss every day. When I was researching definitions for cooperatives, *socialism* was offered as one synonym. Don't get your get your hackles up about that word; I mention it just to point out that cooperative memberships are about communal ownership, not individual privileges. Keep in mind that sharing the work, the investments, and the products is not for everyone. Here are some things to think long and hard about before you decide to get involved in a co-op:

- **Understand all the rules.** If you can't play by them, or you go into it thinking you're going to change them, don't join.

- **Be sure the co-op's message or stated mission makes sense to you** and that you can support it through your products and your actions.

- **Be sure you're group minded.** If you tend to be a loner or are a ramrod entrepreneur with major capitalist intentions, a cooperative is probably not right for you.

- **Be ready to commit long-term.** Once you form something like this, selling, transferring, or quitting will be much more difficult than if you're a stand-alone business or sole proprietorship.

- **Set the ground rules early.** Don't get lost in the great ideas and forget to work with a professional to make sure your membership rules are legal and fair. There are many consultants and trade groups that can help you work out the particulars of your co-op.

## Marketing for Co-ops

All of part III of this book is about marketing, but most of the focus is on marketing individual businesses. Here are some marketing tips to help endear your co-op members to you:

- **Host people.** Let the community truly be involved. Invite schools, community and civic groups, and local government to learn about your program. On-site visits are the best way to stir up interest—and they help with financing.

- **Profile your vendors.** One of your most powerful marketing tools is that you are a local organization. Let people meet your suppliers by posting profiles and color posters of farmers who either sell to the co-op store or grow for the CSA. Ask them to occasionally work on site or demonstrate something unique about their goods. Member profiles work, too, to showcase the different kinds of people who are joining your co-op.

Typically, anyone can shop at a co-op market, but members receive a discount.

- **Give help.** Recipes work wonders for product sales. Ask volunteers to help new shoppers find what they are looking for and recommend companion products—this will also increase sales.

- **Give demos.** Cooking classes on site (check local health department rules first) are a great idea to teach members the difference between couscous and bulgur, for example, and enable them to try before they buy. Or what about hosting a fabulous autumn meal at the farm, so people can revel in the bounty and sign up early for next season?

- **Network.** Make sure someone on your staff is keeping the message out there by leaving the premises once in a while. Join the local organizations that you hope will support you, such as the chamber of commerce, church, or Rotary Club. You can keep your message in front of people just by your good works in other civic pursuits.

- **Have membership drives.** Conduct them at least annually, and offer specials and other features.

- **Keep your members working.** Co-ops were built on the premise of volunteerism, so remind members that they own shares in the business and ask them to help out from time to time. Maybe you need a building painted or some fall clean-up and composting help around the farm. Many cooperatives offer a rate and membership structure that includes a requirement that members work, without pay, a certain number of hours per week or per month.

- **Get noticed.** One of my favorite marketing strategies applies to just about any business and is always free. Promote your business in the media, both locally and nationally. Call radio stations, newspapers, bloggers, and lifestyle magazines; tell them your story and ask if they'd like to do a feature. Having a simple media kit ready is a great way to do this effortlessly. Free press coverage has always worked as well as advertising for me.

- **Be responsive.** The best way to add value is to be valuable. Keep the tried and true, but respond prudently to customer requests and new ideas by staying innovative as much as possible.

## Bring a Cool Head to a Conundrum

A *conundrum* is something that is confusing, mind-boggling, or just plain convoluted. I hope that's not how you are feeling as you read this chapter. It's true, the lines between a food co-op and a CSA are blurred, and that's okay. Really, you're probably already seeing that there are no clear boundaries in the entire hobby farms and local foods movement; they are as intertwined as blackberry brambles in summer.

I don't think this is a problem. Sometimes it makes sense for things to overlap to keep an idea moving forward, combine resources such as advertising and marketing, and create buzzwords and interest that drive consumer purchases.

The only time I worry about the details of all this overlapping is when it causes consumer confusion. And frankly, that happens in this growing niche way too often. You may even have picked up this book in the hopes of finding some clarity about the terms, topics, and trends in hobby farming. I need clarification

> **A media kit,** in its most generic form, is a package of information that you have on hand to give to members of the press or other centers of influence to whom you want to provide a standard message. You can include a few press releases about your recent activities, a one-page fact sheet about your organization, brochures, and business cards for key contacts.

## Local Emphasis, Community Connections

There is a connection from consumer to farmer that Lost River Market & Deli has been able to foster. Brad feels it every day as he drives into the little town of Paoli to work at the store. "It's about community," he says. "Every time there is an *E. coli* scare, people feel they can come here and trust a name and a face they're familiar with."

Brad believes strongly in keeping the produce and other goods as local or regional as possible. When the store opened in October 2007, more than 15 percent of the inventory came from local purveyors. That number is always expected to increase exponentially during the summer months, as the region's produce bounty comes into bloom.

"We spent over $10,000 with 16 local vendors, and that is totally new business for those vendors," Debbie says of those first few winter months. In high summer, "Over half of the case was filled with local produce," Brad says proudly. "But more and more, there are other products from local producers like wildflower honey, maple syrup and sorghum, and black popcorn. The black popcorn has been really popular."

Producer interest in selling to the store has been strong. "In terms of working with local producers, it's been a great core group of five vendors up to 25. This includes Amish producers, and we welcome that sector of the farming population," Brad says.

He continues, "The vision for the store is 80 percent natural, organic, or locally grown with 20 percent commercial product." But he's rational in that desire, knowing that availability and price are top priorities to his members. They've been keeping track of sales from commercial vendors and local purveyors and are pleased with the results so far. Brad says they figure 67 cents of every dollar circulates back to within a 100-mile radius of the store.

"This has a multiplier effect on our local economy here in Orange County," says Brad, and the effect is not strictly in terms of sales. The money also supports wages, the local credit union, regional vendors and suppliers, and even the local churches.

myself some days, and I do this for a living. But that's a good thing; it shows how fast this niche is growing right now.

As you decide on one of the venture ideas explained in this book, including cooperatives, I caution you to make sure you understand the differences between yours and the next program down the road. Certainly, marketing is easier and more sensible that way. But it goes beyond that. When you begin to produce food, you have a responsibility to consumers and to your fellow small farmers. If you have never been part of an industry so integral to human life, this concept is really humbling. For example, if you're not Certified Organic, don't call your product organic, out of deference to those who have spent the time and money to participate in certification programs. Rather, explain why you've chosen not to use the term or the formal program and why this choice makes sense for you and your customers.

When you're starting a cooperative, clarity is really important, especially when you're trying to educate would-be members. Explain to members what ownership they legally have, if any, and what ownership is there just for the produce of a season or of one animal. Don't set people up for disappointment.

You've seen in this chapter how projects in local foods and niche agriculture are tied together. The chance to encourage the development of a cooperative venture to promote health, wellness, education, and individual success is deeply valuable. So is simply creating places that bring people together.

---

## The Lost River Team's Best Practices

- Start the organizational process with a strong commitment and capable leadership.
- Realize that forming a cooperative is going to take time and a lot of work. Be diligent at that process.
- Seek outside resources and the help and advice of professionals who have set up cooperatives before.
- Set funding goals and engage everyone in meeting them.
- Involve local producers whenever you can.
- Bring things to the community that they can't get in large retail stores.
- Make a connection between farm and food.

# Chapter 6

# Agritourism: Recreation as a Business

> **Learning Objectives**
>
> 🌿 Understand what agritourism is.
>
> 🌿 Learn about various types of recreation and tourism on the farm.
>
> 🌿 Meet the people who run Cook's Bison Ranch, a successful agritourism business.
>
> 🌿 Understand the potential drawbacks to living and working at home and hosting visitors as well.
>
> 🌿 Explore some trends in agritourism.
>
> 🌿 Uncover potential business ideas.

## *Cashing In on a Lifestyle as You Live It*

For some, there can be no more perfect life than one in which work, home, and family combine. I live this way, writing, raising cows, and enjoying time with my husband all at the same place, often all in the same day. I would hate to think about driving away each morning to a workplace separate from my home.

Some days I enjoy my routine, and others I just go at my own pace rather than forcing myself into a cookie-cutter day. I work on farm projects, feeding cows, weeding my herb garden before it gets too warm, walking the quiet roads in the lavender light of just-dawn. Some mornings, particularly winter mornings where it stays dark until an unmotivating 8 a.m., I huddle at the sanctity of my desk, catching up on e-mail, maybe doing some creative writing while I'm fresh and sipping coffee from a mug—not a biodegradable paper cup while driving a car. I do love it.

Do you want to work from a home in the country? Would it work for you?

Think again about work and home, and consider for a moment having a business where you not only work at home, live there, and have your family around, but also bring others—strangers—into your sanctuary. Does that idea sound intriguing and even exciting, or is it horrifying? Does inviting the general public to traipse through your gardens, pastures, and country lanes and even lounge on your back porch sound wonderful and inviting, or like a worse infringement on your privacy than an FBI wiretap?

If the idea of combining office, home, family, and customers all in one location sounds like a great way to make money without having to leave home, then agritourism just might be the hobby, foodie, or small farm enterprise for you. But if you're the sort of person who loves your privacy—*a lot*—it may be wise to skip to the next chapter.

## The Pros and Cons of Working from Home

Running a business at home has obvious economic benefits, including savings on fuel, food, clothes, and supplies. From a time-management standpoint, it's also an efficient workplace since you don't spend time commuting to and from another office, and there is no traffic to get stuck in. Your work is always around, which means you can easily break up your day into specific tasks. And there is the freedom that comes from knowing you can run errands, get your hair cut, or simply take a nap if you like.

There are drawbacks that I want to make sure you consider before I explain agritourism in more detail. When I started Aubrey's Natural Meats and began running the business from my farm full time, I was excited and relieved to be working out of my house and to have a single focus, rather than worrying about my full-time job and my new company at the same time. I thought I was ready to work from home. It turns out I wasn't.

My biggest struggles were not about staying on task or trying to keep motivated—although those can be problems for some. Instead, I was lonely. For almost a year after I quit my job and worked on building my company, I felt sort of down. I never attributed it to the change in my lifestyle. Rather, I just assumed I was stressed about making the company work. But over time, I realized I missed the day-to-day interaction with others.

I figured I'd be so busy running the new meat operation that I'd have no time to get lonely. And in fact, I *was* busy. But I felt unplugged from the world, and it affected me for more than a year.

A baby heifer calf enjoys a May day. For many people, a trip to the country to enjoy the sights and sounds of your farm is their idea of a great vacation.

As I became aware of the problem, I started changing my lifestyle. I got involved again in other activities besides raising cattle and selling meat, and I blossomed in happiness like a tulip in April. I'd regained my balance. But the transition to working from home sent me through a surprising low period, emotionally.

Many people report feelings of isolation, loneliness, and boredom, even when they're very busy with work. And these aren't the only pitfalls to think about as you consider running your business on the farm. While it's handy to work right on site, there are inconveniences, too. If you worked in town, you might stop by the bank or the grocery store on your lunch hour or after work. But if you work on the farm, you have to factor into your day the time to "run into town" to access the services and goods you need.

Another work-at-home problem is the one I'm convinced will never go away, particularly if you open the farm to the public. Distractions are endless, and it can be challenging to set a routine and stay productive when people come and go and you've got ten days worth of laundry that you need to get washed.

## Meet Erica and Peter of Cook's Bison Ranch

Northern Indiana is Amish and Mennonite country. A host of knick-knack-laden flea markets, restaurants boasting hearty home cooking, and handmade wooden crafts draw thousands of tourists each year to the slightly rolling hills and peaceful prairies. Peter and Erica Cook are not Amish, but they, too, benefit from the region's tourist attractions. The Cooks operate a bison ranch.

*Bison,* you say? In *Indiana?* Yes, bison in Indiana—but that wasn't the original intent of the farm or even the lifelong dream of the proprietors. "I had no farm background at all; I grew up in South Bend," admits Erica. "I was only 18 when we got married, I guess I just followed Peter with what he wanted to do." Peter was from the area, and the farm has been in his family since his grandfather homesteaded the place.

Peter didn't actually grow up on the farm. But in 1998, a series of circumstances led the Cooks to add buffalo and start an agriculture business.

"Peter graduated college that year, and the farm ground was coming out of set-aside [a conservation program in which the land is not farmed while enrolled in the program] at the same time," Erica recalls. Peter fell in love with buffalo on trips out West during his college years, and his family was looking for a way to use the land.

"It was a good fit," Erica says. "The bison were something unique, we liked the health of the meat, and there was a growing

One possible solution to dealing with distractions is hiring staff to do some of the chores, such as mowing the lawn and cleaning the house. For example, I have a housekeeper who cleans everything, including the inside of the refrigerator, and even does the laundry. If you're not looking to spend extra money, try challenging yourself to work on those home-living chores just once or twice a day.

I hope I haven't scared you away from working from home or inviting folks out to the farm. Once I got past some of my initial transition woes, I found my work-at-home lifestyle so rewarding that if I skip a day in my home office, I miss it. But this book is all about providing you with the information you need to make a smooth transition, and we're discussing the exciting and the mundane, the motivating and the potentially discouraging. So, if you still think this adventure is for you, read on. You're likely a great candidate for the small farm business niche called *agritourism.*

market for them. We like the historic aspect of raising buffalo and that they were a niche market." Initially, the Cooks planned to sell breeding stock because at the time calves were worth upward of $2,000 each. But eventually calf prices dropped drastically. "So we decided to switch to selling meat," she says.

Erica eventually earned her teaching degree, and that focus on education has proved to be useful in their agritourism venture. To supplement meat sales to restaurants and distributors, the Cooks decided to capitalize on their novelty and their location. So they added farm tours.

Before long, much of the family was involved in the operation. Erica values the opportunity to be at home with her children, Lucy and Levi, and she schedules the tours and organizes farm events. Peter, who had previously been in the restaurant business, now works full-time managing the bison herd. Suzanne Garza, Peter's sister, manages the office, and another sister, Annette Moore, and her husband, Jason, are also involved in the business.

By adding tours, Erica believes the family is able to establish a relationship with customers that will lead to more sales later. "The meat portion is still 95 percent of our income," Erica says. "The tours seem to help our meat sales because people get a taste and it is cooked properly."

## What Is Agritourism?

Agritourism is one of those made-up words that's now common in rural-speak. The term's origins seem to be unknown, and it probably has existed for only ten or fifteen years—or about as long as the trend of visiting farms has been exploding. Of course, many "gurus" of agribusiness have offered formal definitions for a word that was invented by the industry. Generally, agritourism means what it sounds like—the opportunity to tour or visit a destination that is related to agriculture. Consulting firms and trade groups have broadened that general meaning to include a business that conducts tours or visits and adds an educational component to teach visitors about farming, rural life, or local foods.

Agritourism can also easily mean an agriculturally themed vacation—and therefore, my very favorite type of vacation. Visiting wineries and wine tasting is certainly a popular form of agritourism.

Peter and Erica Cook and their children, Lucy and Levi, in a pasture with the bison.

Beyond the educational and the agricultural, agritourism can also mean agri-entertainment. Again, like it (sort of) sounds, this just means fun activities within an agricultural setting. Many farms offer the agri-entertainment component along with their agribusiness and food and education to lure visitors. A perfect example is a small farm orchard that presses and sells cider, allows people to try samples and pick apples, and offers fun diversions such as hay-mound "mountains" and corn mazes.

If you're a **recent transplant to rural living,** you may be in the best position to set up an agritourism venture that will interest consumers like you!

In browsing for formal definitions, I also found one as simple as describing agritourism as "paid farm visits." That one sounds a bit clinical, don't you think? Yet it, too, is accurate (albeit a tad short and terse) because agritourism is a business model, and it does imply that someone is paying a fee to visit the farm and engage in the activities the producer has for sale.

Agritourism at its most basic level combines education and entertainment in the context of agriculture and rural living. Here are some categories you might see:

- On-farm market
- U-pick farm or orchard
- Farm and livestock tours
- Agri-entertainment
- Winery
- Pay-hunting or pay-fishing
- Guest ranch
- Trail riding and rural scenic tours
- Bed and breakfast
- Creating something from farm products that visitors can take home (soy candles, for example)

There are plenty of definitions and categories, but the concept, only recently recognized in both agriculture and tourism as a separate niche, is still trying to find its direction. There is still a lot of room in this industry for paving one's own way. I think that's why it makes a great fit for newbie farmers. Conventions, rules, and traditions pretty much don't exist, which means new farmers can "get it right" no matter what their unique perspective is.

## Agritourism Trends

As I mentioned in chapter 1, statistics on the agritourism industry are muddled. Many are collected as part of a region's general tourism numbers and are not differentiated. In some places, statistics are not kept for this niche. Further, statistics seem to be strictly regional, and there is no clear resource for packaging facts on the entire industry nationwide.

What we do know is that in most states, agritourism is growing and has been for more than ten years. Recognizing these businesses as profitable ventures that build better rural communities, state and local government agencies have put money toward these initiatives, which have become increasingly better organized and better funded. Many states are partnering departments of agriculture with departments of tourism and creating web sites, guidebooks, and recommendations for visitors. Promoting and supporting agricultural vacations is gaining attention and funding, which is a benefit for every small operation looking to start a new project.

## Business Ideas

Many small farms combine agritourism with another type of business model, such as selling at a farmers market or delivering product to stores or directly to customers' homes and businesses. Even CSAs and food cooperatives have an agritourism element to them when you consider the subscriber or member's recreational component of visiting the farm to pick up their weekly bag of food or helping out at the co-op store.

While the annual percentage growth is not as strong or as lightning-fast as other alternative agriculture niches such as organic foods, agritourism has grown steadily. In many instances, farming organically or with a natural focus works well with agritourism. **This fosters crop diversity.** Increasing acreage of traditional row crops like corn and soybeans is being replaced with vegetables, fruit, and other food crops that are conducive to farm stands and U-pick businesses—even in corn-belt states like Indiana.

By combining business types, you capitalize on a couple of different avenues for commerce and give your buyers options about where they want to buy your products—which, hopefully, boosts sales. The combination of hosting people on the farm and conducting off-farm business works for many small farms, but keep in mind that it can be really hard to manage on a shoestring budget with a small staff. Plan carefully, and try not to have your products overlap; keeping the offerings on the farm and through markets or stores unique.

As I've already mentioned, there are lots of options for agritourism business. This section outlines some additional ideas in combinations you may not have considered. Try a few things, and don't be afraid to change elements from year to year or season to season to keep people visiting for both a favorite activity and to try something new.

## Just Tourism

Pure agritourism, where you establish a business exclusively with the idea that what people are buying is a visit to the farm, is possible. Many people find that this business model works for them. Ideas include:

- Bed-and-breakfast or guest cottages
- Outdoor activities such as hiking, photography, biking, picnicking, and horseback riding
- Petting zoo or animal tours
- Demonstrating how a farm operates
- Hosting educational seminars

## Agritourism Around a Theme

One of the agritourism ideas that can be both unique and effective is creating a farm experience around a theme. Take animals, for example. Sure, you can host a petting zoo, but why not make it a more detailed experience by teaching people about the lives of farm animals in all their cycles?

Erica and Lucy Cook join a visitor on the tour wagon.

# Tours for All Ages

Tours at Cook's Bison Ranch are the biggest draw for visitors. Erica and Peter want to make them fun, of course, but they also want to help consumers better understand bison and encourage them to choose bison meat. "A lot of people thought that bison were extinct or endangered," Erica says, so they avoided the meat. When they realize that bison are farmed like cattle, their idea changes.

Many tours are targeted at elementary school children. "We start the little kids out by trying to educate them about the animals and the meat. We hope they'll tell their parents and get them on board. Then, when the kids grow up, they'll be more comfortable buying bison," Erica explains.

"We first started kind of generically with a wagon that was in the barn, and we fixed it up with new tires," Erica says, adding that it seated only 15 people. "We had just one class at a time at first. We would do lessons and play games and talk basic facts about animals and go on a ride, then have a meat sample."

The tours originally focused exclusively on school-age children. But eventually, they became so popular that bus tour groups of adults began to visit as well. Suddenly, the Cooks had to modify their original modest plans.

Since 2000, the operation has added more wagons and can accommodate up to 100 people at a time. The Cooks purchased a concession trailer so they could offer meals at the farm. Because of the various age groups that come for tours, Erica divides the tour stops into stations so she can address different interests.

With large bus tours becoming common, they now cater to their meal needs by hosting chuckwagon-style meals for 35 to 50. "Some visitors walk in, but most are done by reservations," Erica says. "With my little ones at home it just works better that way!"

To keep customers interested and comfortable year-round, the Cooks have added a retail gift shop and permanent restrooms at the farm. "There's usually someone there every day, so at least people who stop in have a person to talk to," Erica says. "We could have tours every day if we hired someone else, but we think it's hard to grow the business without the personal touch. It's better for the customers if they get to know us."

Consider hosting people to see and even help with calving or lambing, or to learn about making hay and managing pastures. If you work with a processed food, such as milk or cheese, guide them through the entire effort, from milking to finished product, and send them home with a treat that is planned and purchased in advance.

Agritourism centered on food is growing in popularity, too. Try offering a special harvest meal using all local (or better yet, your own) foodstuffs, and explain the life cycle of the vegetables and proteins. You combine entertainment and education in your beautiful setting—and, of course, charge for the meal.

Your marketing focus is where the best potential customers can be found. Market your agritourism ideas to schools if you like the animal theme and to gourmet clubs and even restaurants if you prefer the food and farm meal idea. If you've got a beautiful herb garden, why not host master gardeners, garden clubs, and even retirees or members of an assisted living community to walk the garden and craft a small herbal souvenir, such as a lavender sachet or a freshly collected bag of mint to steep for tea?

Events like these can be announced well in advance and hosted several times a year to keep repeat visitors interested.

## Agritourism Around a Season

Agriculture revolves around the seasons, and once you move to the country, your awareness of the seasons deepens. Why not share that awareness with visitors? Tasks are different on the farm every time the weather changes, so consider hosting events or even a yearlong series of events that showcase the seasons on the farm. For ideas, look no further than what you'll already be doing and decide whether you can make time to host paying guests in the midst of your activities.

In the spring, there's preparing your seedbeds and planting as well as animal birthing. Summer brings many fruits and vegetables to pick and flowers in bloom. Fall yields the last of your produce and is the time to can, freeze, and preserve your summer crops. Even in winter there are tasks such as livestock chores.

Seasonal ideas can also be found in aspects of the rural lifestyle that are unrelated to agricultural production. For-pay fishing and hunting are great options if you have an expanse of privately owned woodlands and prairie grass. Just be sure to investigate state and county permit requirements to offer groups the

If you're going to create an agritourism business around an existing idea, such as hay rides, corn mazes, and apple cider, **merchandize it a bit with your own flair for the unique and untried.** Being personal and different while offering something comforting and homey is the balance that wins visitors.

## Agritourism Seasons at Cook's

At Cook's Bison Ranch, tours are offered May through October, when the weather in Indiana is most pleasant. "Calves are born in May, so that's a nice starting point for the season," Erica says. While that's a great time for visitors, finding a balance between hosting guests and doing the chores required in the summer months (such as making hay) is challenging.

In addition to bus groups and seniors, the Cooks host summer camps, YMCA groups, and libraries that conduct summer learning programs. "It's hard to say which groups like the bison tours most. All age groups have a good time. The city kids seem to be most interested and more excited and have the most questions," Erica laughs.

Still, the different times of the year bring different groups with different interests. Erica caters to each visitor as much as she can. "We adapt to different groups. I can tell when a group gets there what they are going to be like, so I have a couple of different options." Erica says she can tell if a group of youngsters will be high energy and need activities to keep them busy. If so, she focuses more on animal interaction and less on "lectures."

opportunity to hunt. Also look into what (if any) game seasons and licenses will be required of you and your hunting guests. In the winter, walks across frozen streams may interest guests. Cross-country skiing and snowmobiling are also good ideas. In the spring, mushroom hunting is enticing to a lot of gourmands. And birdwatching is a huge industry.

As a nice plus, perhaps your guests can help you out. Offering working farm experiences could reduce some of your workload during busy times and help you add labor that's not ordinarily around the farm.

# Partnering with Other Farms and Communities

*Co-marketing*—that is, sharing marketing costs, plans, and workload with another, noncompetitive farm venture—saves money, provides more options for visitors, and gives you an instant customer list when you pull from existing events or neighboring agritourism ventures. As you'll see in chapter 10, growing your customer database is your opportunity to build connections that translate into repeat business.

# Happy Trails

Have you ever taken a bus tour of wine country or picked up a free map at a rest stop and driven from winery to winery with your friends? It's fun, isn't it? Wine trails are some of the best-known partnership initiatives in agritourism today. They work rather well, in part because they acquaint visitors with several different stops on a logical, convenient path that makes sense to dedicate an entire day or an afternoon to exploring.

The concept of a wine trail is transferable to any group of agritourism ventures in a geographic area. You don't all have to offer the same agricultural product, either—wine is simply an example. In fact, partnering with several firms that operate different types of businesses could better encourage tourists not to miss any stops along the way.

Your arrangement with other agritourism ventures can be formal. For example, you could put together a county or regional association and work to better the entire area. Or it could be loosely operated from season to season.

To get started, you need to learn enough about your neighbors that you would be comfortable recommending them to your valued visitors. Marketing can include printed maps and flyers, or start as small as offering "the trail" on each firm's web site with a map and directions. To kick it off, why not host a special theme event? Consider punch cards that are marked at each location and earn the visitor a little gift at the end if they get to every stop.

An aerial view of the farm and the bison on pasture.

## Extras at Cook's That Visitors Like

Admittedly, seeing a bison roaming is something most Midwesterners aren't accustomed to. But it takes a bit more to keep folks coming back to the farm year after year. "We do a lot of things that the guests like," Erica says. "We offer door prizes, we send postcards and thank-you cards after all visits. The host or driver of a group gets a gift, and teachers get handouts for them to expand on. They really appreciate it."

Making customers happy is a full-time job for the Cooks, but they feel it's worth the effort. "A lot of it is our hospitality and the way they're taken care of," Erica says proudly. "The farm makes many people recall their own life on a farm. Coming here brings back a lot of memories for the seniors."

The efforts pay off in lots of happy visitors. And Erica says happy visitors make her and Peter glad that they've kept the bison tours going. "It is a great experience to see how happy people are when they leave," Erica says sincerely. "It's no fun just keeping the farm all to yourself!"

## Fairs and Festivals

To draw visitors to the farm, your partnership doesn't have to be a formal arrangement with another group. Try planning a big annual event, an open house or a themed tour around the time of a community activity that's already well established. Ideas include the county fair, your town's heritage days, fall festivals such as Labor Day and harvest fairs, and even wintertime favorites such as caroling and Christmas walks.

You don't have to be a member of those organizations to piggyback off the crowd that is already used to visiting during the day or over the weekend. Make your activity related but unique, and make sure it's convenient and simple for visitors to find you from the festivities they're used to attending. Post flyers and road signs both in town and along the route.

## Tips for Keeping It **Your** Farm

If you've talked to other agritourism business owners who host special events, you may think I should have titled this section "How to Keep Your Sanity." But that's not something I'm here to advise you about. To put it another way, no matter how excited you are about hosting people on the farm, at some point you're going to want to close the proverbial barn door and be left alone. That's when firm yet polite management of your time, property, and personal possessions becomes a delicate but essential task in creating and maintaining an agritourism venture.

The best defense against stop-bys when you are not open for business is to plan in advance when you want to be open and stick to it. Here are some more specific tips:

- Manage your farm hours and make sure they are posted everywhere, including the Internet, your farm's gate, flyers, your business cards, and even little stickers or magnets you put in customers' shopping bags.

- If you change hours or close for a season, e-mail your customer list and any person or group who might be sending you business.

- Post signs on the farm about where visitors can and cannot wander.

- Post signs warning people to be careful around potentially dangerous areas, such as electrified fences.

- During especially busy times, keep a few extra friends and family around to direct visitors and gently remind them to stay in allowed areas.

- Post signs that clearly say where parking is allowed.

- Offer outdoor, portable restrooms and hand-washing facilities.

- Ask guests to stay out of your home. (Post a sign or just lock the door, if it comes to that.)

- Realize that people will just stop by, and keep the farm ready (and secured, if you're away from home) for visitors you weren't expecting.

## Special Considerations for Agritourism

All agriculture concerns have unique considerations to examine. Liability is one of those, and it's discussed in detail in chapter 4. Agritourism carries the highest levels of liability because you're regularly inviting people to the farm who may not understand farm dangers. If you're working with animals or equipment, your potential risks increase. And if you add in selling meats, cheeses, or cider (all foods that departments of health consider potentially hazardous), you score one of the largest potential liabilities in the farming industry.

## Potential Drawbacks and Liability

Sharing the farm may be one of the Cooks' goals, but that still means they are literally sharing their home farm with the general public. Erica says they've been fortunate that there have been no problems in the nearly ten years since they've been giving tours, but she recommends that others considering agritourism plan ahead and protect themselves.

"You definitely need to contact an insurance agency and have a certain type of insurance coverage for giving tours on the farm," Erica recommends. She says some friends in the industry have struggled to get insurance for their agritourism enterprise or ended up with insurance premiums that were ridiculously high. She advises that farms look to insurance providers who know agriculture and understand what will actually be happening on the farm. The Cooks use Farm Bureau Insurance, for example, and have been satisfied with the premiums and the coverage.

Erica says farms can expect their insurance agency to come out periodically and inspect the entire facility. In her case, they look over the wagons and fencing and make sure they meet safety standards. Erica thinks this relationship offers a measure of security they can't do without.

Yet even before you get the right coverage, knowing if you're the right person for the job is the first consideration. Erica admits there are drawbacks to hosting people on the farm. "Anyone considering agritourism needs to have the right personality and a passion for people. We have something we love here, and we just don't feel right keeping it to ourselves."

The time commitment is another major consideration, particularly the unexpected tasks. "A lot of times when you're not expecting to do any tours, someone will just show up and they've driven a couple of hours; you can't just turn them away," Erica explains. "You need to be willing to sacrifice your own time."

My advice is to get serious about setting up an LLC for your agritourism venture. Talk to your lawyer and accountant about the business. Also, your insurance agent will likely recommend more coverage and even an umbrella policy larger than the amount of coverage you already have for the farm. Take liability issues very seriously for agritourism and plan accordingly.

Another consideration that can be challenging for agritourism is seasonality. Many successful agritourism ventures have busy seasons in the fall and summer and very limited business in the winter months. If that seasonal model is your goal, consider how much work and return you'll achieve for being open just several months of the year. Can you make enough money with a seasonal model to justify starting the business? If not, will a year-round agritourism operation work in your area, and do you have enough activities to keep people interested when it's cold outside?

Finally, ask yourself if you can handle it all. Working exclusively from home can seem like a labor savings since you won't necessarily need employees to deliver products or sell at a farmers market. But if you do get good crowds or host meals or overnight guests, what will that do to your workload? Can you market your business and serve those who are visiting all at the same time? I like the idea of family volunteers during busy times, but volunteerism often exists only when it is handy for the volunteer. Think about temporary staffing for special events.

## Places for Help, Support, and More Ideas

Agritourism is a very personal business model because you're hosting visitors to your farm, and they are looking for an experience that reflects the lifestyle you lead.

However, if you're not sure how to identify that special angle only you can offer, there are agritourism consultants. A web search for "agritourism consultant" will yield a number of resources you can call and interview. Most of these professionals will come to your farm, assess your situation, and then write or present a plan of action that even includes follow-up consulting and coaching. Web design and training is often included or available as part of a package.

Your local land grant university is most likely working on agritourism with other small farmers and hobbyists. Contact any university department that offers rural development help or new ventures assistance. Many universities offer free resources to read and even in-person consults to get you started. This type of assistance is usually free. Some universities also have free podcasts that you can download or webinars featuring other entrepreneurs.

Call your state's department of agriculture, ask if they are working with any agritourism initiatives, and find out what is available to help you get started. You might uncover matching grants and training sessions, or at the minimum you could add your enterprise to a list or web site that mentions agritourism stops around the state.

Finally, check the resources section of this book for ideas and web sites to explore for more help getting started with any hobby idea in the book.

## *Marketing the Bison Ranch as a Destination*

Marketing for agritourism is as time-consuming as marketing for any other type of small farm hobby or food venture. Erica recommends coordinating some of the marketing efforts with other organizations that have the same goals, such as tourism groups in your area. "We partner with the local convention and visitors bureau, and they send a lot of business our way. They really help us cut our advertising costs," she says.

Because of the Bison Ranch's proximity to the major Amish settlement in the Midwest, motor coaches bearing tourists flood the region all summer long. "We started with a couple of hundred people a year, and now we're over 4,000," says Erica.

The Cooks use a variety of other marketing techniques to keep the locals interested. Every spring they host Calf Days, an event that includes free tours and activities. They even have Native American dancers and a woman who spins buffalo hair into yarn. Last year they drew 1,000 people during Calf Days alone!

To get ideas for their event, Erica calls the local convention and visitors bureau (CVB) and they help her network with other businesses that can either serve as vendors to the Cooks or offer additional entertainment during the event. Networking contacts provided by the CVB have helped Erica broaden the quality of Calf Days and reduce costs. "A lot of times people approach us first; we just try to make Calf Days unique."

The Cooks also maintain an active membership in the National Bison Association, which enables them to network and connect with breeders and interested people all over the United States.

## Be a Joiner

One of the top things I recommend to new small-farm entrepreneurs is to get involved with local associations and trade groups. Taking time out from your new adventure to join in might seem counterproductive, but there are many reasons to devote a few days a month to learning and networking.

First, learning and networking are the easiest, cheapest ways to get training and avoid expensive rookie mistakes. If I did not believe in finding out the best way to do something from someone who's been through it, I wouldn't be writing this book. Nothing replaces practical experience. It's the most valuable research entrepreneurs can get.

Bentley, an orphan calf, was donated to West Texas A&M University. He will be trained to be their mascot. That's the kind of publicity you just can't buy.

The second reason for getting involved goes back to my earlier concern that some people working from home suffer from a little bit of depression or the gloomy, let-down feeling that occurs the day after Christmas or the day you return from a long-anticipated Caribbean vacation. Once the newness wears off, working from home can get lonely; there is no watercooler or break room in which to chitchat about sports scores and vent about the boss.

Now that you've moved to the farm, you may have less in common with your old friends from town as well. I went through a period of feeling like a housewife because my girlfriends working in town were dressing up every day to go to the office and I was walking across the hallway in my bathrobe. If you know you have an event coming up where you can celebrate and commiserate with like-minded people, it will really ease the transition for you.

Finally, consider getting involved in your community. If you are new to the area, meeting your neighbors is as easy as going where they go to congregate and joining organizations that serve the community. If you find that services are missing from the community or should be there to help serve your new business, work to help see them offered. On a winter's day when no one has stopped by for business, feeling needed is a real pick-me-up.

## Erica's Best Practices

- Hire good staff, whether it's family or outside help.
- Make personal contact with all customers a priority.
- Be able to adapt your farm activities to different age groups.
- Be flexible in your programs; have options and learn to read group dynamics.
- Understand that agritourism requires a love of people and a major sacrifice of your own time and resources.

# Part Three

# Selling, Marketing, and Prospecting

# Chapter 7

# Wholesale and Retail Pricing Strategies

---

### Learning Objectives

- 🌿 Understand the differences between wholesale and retail price.
- 🌿 Explore various pricing strategies for small businesses.
- 🌿 Consider the pros and cons of selling retail versus wholesale, and learn how to use them together.
- 🌿 Get some pricing strategy examples from small farmer and college instructor Rebecca M. Terk.
- 🌿 Understand how to evaluate and change prices, including pricing for sales and discounts.
- 🌿 Consider how much pricing information to publish and how much to withhold.

---

## *Meeting Your Public and Making Money*

It will soon be time to emerge from the office and either invite folks out to the farm or bring your homegrown, homemade bounty to them. In this third part of the book, I introduce you to the next step in setting up a hobby farm or local foods business: marketing and selling. The four chapters in this part cover pricing, marketing, selecting sales venues, prospecting, and creating a customer database. I'll point you to some free marketing ideas and ways to keep those hard-earned customers coming back year after year.

Business planning and market research—what you learned about in part two—are important first steps, but you'll be using marketing and selling strategies every day. Once you've started marketing yourself, you'll never stop, not for one day or one minute. Don't get exhausted and head back to your cubicle; many of the concepts I discuss in this part are just plain common sense and will soon become second nature. Besides, you may find marketing and selling to be a lot of fun!

The challenging part about marketing and selling when you live on a farm is balance. If you have been a marketing executive or a retail shop owner, great. That experience puts you ahead of the game when you start this new endeavor. But you'll probably find that keeping up with marketing and selling is much more difficult when you are also busy raising animals or managing crops. As always, I advocate having a well-considered plan before you get started, so you're ready to stay on task and keep the stress level a little lower.

## The Five Pillars of Small Farm Business

Here in part three, we'll be working with a concept I call the Five Pillars of Small Farm Business. Taken as a whole, these are the elements necessary to make money and stay in business, no matter how small your business is:

- Effective pricing
- Marketing
- Selecting selling venues
- Prospecting
- Retaining customers

Pricing is not only deciding how much to charge but also what type of selling you'll be doing—retail, wholesale, or both. Marketing is broad and includes everything you do to get your name out there in front of customers. In selecting a selling venue, you'll choose from a variety of places that lend themselves well to rural businesses.

Prospecting for customers is one of my personal favorite topics and favorite business activities, and I've devoted an entire chapter to the art and best practices in the chase for customers. Finally, retaining customers is key to making sure that your prospecting efforts bear fruit year after year.

Each of these topics is covered in the next few chapters. I'll start here with effective pricing to establish the base of your commerce.

A vendor prepares fresh produce for a morning market.

## So What *Is* the Price?

Price is simply the value directly associated with an item or a service. Price represents perceived value, usually in terms of money. Price can also be calculated in terms of worth, which is often the case when goods or services are bartered. Of course, price must eventually be mutually agreed upon between buyer and seller, or the sale never happens.

We inherently understand the price of something most of the time without even thinking about the word. Price and its related concept, value, are at the heart of every transaction. So if we know what price means, then why devote an entire chapter to a concept we use every day? Well, do you know *why* an item is priced at a certain level? When you buy a car or a house or cup of coffee, do you understand the cost of the inputs needed to get the final retail item to you? Do you understand the distribution chain, from raw materials to production to processing to transportation to marketing to packaging to finished goods? Really, why does stuff cost what it costs?

There's also a component of cost called *markup* or *profit* that has nothing to do with the cost of raw materials or labor or doing business. But that's the component that's essential to making money.

For most of the things we buy, we have absolutely no idea what it cost to get it to us, nor do we need to care. But turn the tables and start a rural small business or a local foods venture and you'll suddenly realize that you *must* know all these elements of price from farm to table. You are the producer, purveyor, and marketer, and you've got to know why stuff costs what it costs.

There is something pure in this concept that I like; it is consistent throughout the local foods movement. When the producer is involved in the entire process, he or she can explain that process to the consumer, thereby linking the chain and creating more interconnection from gate to plate.

When you set prices, you have to take a lot of things into consideration and make many choices. And you'll gain expertise with pricing strategies almost immediately as you see customer and peer responses to prices in the marketplace.

Sometimes entrepreneurs tell me that all pricing amounts to is setting the number as high as they can hope to get. Indeed, there is some truth to that, but price isn't always about getting the highest amount up front. It's more about setting a figure that not only makes you money but also communicates value to your customers so that they will come back again and again. You also want them to communicate that value to their peers. If you simply aim for the sky and hope to get it, you might—or you just might push away a whole lot of great customers and friends. There is a distinction between pricing too high and pricing too low, and it requires finesse, experience, and a measured understanding of your market. Understanding where you are on the high-low spectrum involves taking time to pay attention to the market and constant diligence to understand how to create a price that offers both profitability and value.

> It's important to know why you're pricing an item the way you do, but I think it's also your job to **explain this value and the cost of the materials you worked with to the consumer.**

This chapter focuses on specific ideas and concepts, but the first item of business is to think about the price of your items more holistically. Think about the true worth and value of what you are creating. The easiest way to examine price holistically is to start with yourself. If you were the buyer, what would you give for the items? If it's helpful, think in terms of a value price, a higher price, and an expensive price. Where did you get those figures, and how do those numbers compare with the prices of similar goods?

The prices you come up with may be too high if you think, "I can't believe people would pay *that!*" Likewise, if your first few customers are saying, "Wow! All this and it's so darn cheap!" perhaps you'll want to raise prices a bit.

## Meet Rebecca M. Terk, South Dakota Vegetable Grower and Farmers Market President

Everything happens for a reason. Rebecca M. Terk's journey to becoming a small farmer involves a lot of twists and turns, but she seems to have ended up where she was meant to be. Rebecca was born in Vermont and is now an English instructor at the University of South Dakota. She holds two master's degrees, but neither of them is directly related to her small business in agriculture. While her formal education has been in English and history, Rebecca's lifestyle training, so to speak, has been in rural living and growing vegetable crops.

Rebecca moved to Vermillion, South Dakota, at age 20 to finish her bachelor's degree, then went to the University of Wisconsin at Madison to start her on her master's. She liked the area well enough, but found the Big Ten college not to her liking. So instead of school, she took a job at a CSA farm and spent a season and a half getting her feet wet and her hands dirty in local food production.

After that, Rebecca briefly moved back to South Dakota, then Vermont, and worked at an organic market, where she was a produce manager. "I got experience working with local growers and with negotiating prices, but financially, there just wasn't enough there," she recalls. The work was just another component of her education in direct-to-consumer agriculture retailing.

## Why Is Pricing So Important?

It may seem obvious that how you price an item is directly related to how much money you make or lose. I posed the question anyway because I've spoken to far too many new entrepreneurs who either copied their neighbor's pricing or simply said something along the lines of, "We'll figure it out as we go along." That attitude is dangerous and will earn you more work later, plus more customer complaints. Consider these reasons for putting a lot of thought into setting prices:

Heading back to South Dakota, and now married, Rebecca was able to complete her master's degrees in English and history while working and teaching on the Rosebud Indian Reservation. In 2002, she gave birth to her son, Martin.

Rebecca's growing interest in local food production sprouted from her love of fresh food. "I don't have a farm background, but my mother always had vegetable gardens and I knew a lot of kids whose parents were dairy farmers back in Vermont," she says. Moving from the Northeast, where small farms and farmers markets are common (Rebecca says, "If you slap a Made in Vermont label on it, the product sells pretty well"), to the upper Midwest's corn and soybean country, Rebecca realized her food choices were quite different than what she valued as a kid. Despite the breadbasket feel of the Dakotas, the grocery stores just didn't have the produce she longed for. "I was kind of shocked and dismayed at the whole selection," she admits. "Growing up in an area with a lot of fresh, high-quality food, it just wasn't as easily had in South Dakota. I started growing my own in part because I wanted to be able to eat better."

Back in Vermillion to teach, Rebecca also started working at a small organic farm and greenhouse and did some selling for the owner at the local farmers market. She had her own garden, and by 2000, she was selling a bit of her own produce at the market. "The whole consumer-to-producer thing, that's really me!" she says.

- It makes your management and accounting easier.

- You avoid customer confusion, complaints, and problems.

- Having a pricing structure is essential to figure out your break-even point right from the beginning.

- A good pricing structure provides a solid model for tracking what sells and what doesn't, and gives you an idea of why.

- A solid pricing model gives you a good way to figure out whether your products are cheap, expensive, or competitively priced for your market.

Rebecca Terk poses with freshly harvested beets in her garden.

# Wholesale Pricing

Wholesale pricing, by my definition, is the price to which the final retail markup has not been added. In the cow business, we say, "There is still some trade left" in the price.

This type of pricing is typically used when you sell your product to one buyer and it then goes through a series of transactions that leads to a retailer or a restaurant offering the product to the end consumer. You can either sell the product at wholesale directly to the restaurant or retail store, or you can sell through a distributor who buys the product from you at a wholesale price and then sells it at a marked-up wholesale price to the retailer or restaurant. Farmers markets and other direct-to-consumer transactions are almost exclusively considered retail, so I'll cover them in the next section.

Many hobby farms and food businesses sell ready-to-eat, finished foods or goods, but that doesn't have to be the case if you've got an idea that requires further packaging, processing, or finishing. Raw goods may also be sold at wholesale—for example, you may sell the butter that another producer uses to make cookies, that they then sell to a gourmet shop.

## A Tale of Meatloaf and Chili ,

The year I started Aubrey's Natural Meats, I intended to offer my product direct to consumers through farmers markets and delivery and sell direct to restaurants. I also decided to have a retail space. Not wanting to lease an entire store, I ended up leasing space in a gourmet deli, dry goods shop, and high-end wine store. It seemed like a great place to sell my attractive cuts of beef and pork. What happened surprised me, and taught me that the more ready-to-go an item is, the more you can get for it.

I thought my meat cuts would sell, and they did—but only a few. Within days I realized that my lovely steaks and ultra-lean ground beef were going to waste as people came by looking for something to eat immediately. Not knowing whether it would make sense, I quickly called my mom and asked for her recipes for meatloaf and chili. I got a warehouse club card and sent my employee to the store for seasonings, beans, tomato juice, canned tomatoes, and take-out containers. Meanwhile, I tried to perfect the recipes for a food service scale.

In a matter of days, sales picked up for me and product loss was down, but I had turned into a mostly ready-to-eat vendor and not a custom natural meats shop. The labor needs grew as someone had to get the fixin's to make the ready-to-eat items and prepare them in the kitchen while managing the counter. Also, inventory space was harder to come by with bigger containers and more products.

Eventually, I realized that ready-to-eat wasn't for me, and neither was the retail store. While I could earn more money with ready-to-go products, it involved more work and more management. I left the gourmet shop after four months. And that's my meatloaf and chili story.

## How Much Less Is Wholesale Pricing?

The difference between the wholesale price and the retail price of something varies by industry, region, product, and even by customer. One way of looking at it is that the wholesale price depends on how many hands will touch the product after yours; usually, the more hands involved in the wholesale-to-retail chain, the less you, the initial producer, will receive. For example, if you sell to a distributor, they'll likely want to pay less—say, 10 to 40 percent less—than if you sell to a restaurant or a retail location. That's because the distributor must leave some room for its own markup.

## Overworked but Finding a Balance

Once she set her mind to having her own farm business, Rebecca realized she'd bitten off quite a large piece of American dream pie. Trying to balance teaching and parenting meant that the first couple of years had their challenges.

In 2000, Rebecca leased a piece of land in town to farm and sold her produce at the Vermillion Area Farmers Market. The money she made selling vegetables that first season was just an add-on to her main income as an instructor. But the next spring, Rebecca realized that she might have only one class to teach that semester. She knew that wouldn't be enough for her family to get by, so she decided to start her own CSA.

"I thought, 'Well, I guess I better just dive right in!' So I put together this brochure and started looking for members to sign up," she recalls. Her goal was ten members, and she met it. "I was surprised people would give me all this money up front," she says. The initial subscription price was $350 a year for produce delivered weekly most of the summer and fall.

Just when she thought she had a good thing planned, Rebecca was horrified to discover that the land she was renting had been sold just weeks before planting season. Scrambling and stressed, she was able to lease land at the organic farm and greenhouse business where she used to work, which was now defunct.

The amount you can charge for your wholesale goods also depends on how complete and ready-to-use the item is when you sell it. The more ready-to-use the item is, the more you can charge. So if you decide to sell raw materials, your price will be considerably less than if you sell a finished item that is packaged, labeled, and ready for use.

## Working with Distributors

You can sell wholesale directly to the retailer or restaurant, or you can sell to a distributor company that will do the selling for you. The distributor is not really your agent, though; it represents itself and simply offers brands (like yours) that it buys at the source. Distributors will pay you directly and in full for the item you sell, and you'll receive no additional cash or benefits from it when it sells to the end customer.

Acquiring a distributor is at once simple and challenging, depending on your situation and the interest at the distributor level for your products. I worked with two distributors, one in my home area and one about four

Just about the time she was gearing up, she found out that she was able to get a full class load from the University of South Dakota. Now she had a full-time job as a teacher, a part-time job as a farmers market vendor, and a full-time job operating her own CSA! "I was trying to do this all by myself," Rebecca recalls with a sigh of exhaustion. "I guess I was slightly younger then, and I didn't know what I was getting into."

That first year, she grew more varieties of vegetables than she does now—between 50 and 60 items (she's now pared back to about 40 of her best-selling and most requested items). Though her first season had its rough days (and plenty of exercise), she has found a nice balance of work, work, and play that she is comfortable with. She gardens on one acre of land owned by her partner, Harry Scholten, and teaches an 80 percent schedule at the university. Her classes are exclusively online, which enables her to work around her agrarian pursuits from a laptop at home or in the field. The CSA takes much of her produce, and she still sells at the Vermillion Area Farmers Market all summer.

Today, Rebecca estimates her income from the food business at around 15 to 20 percent of her take-home income.

hours south of me that distributed in southern Indiana. I made the deal simply by calling and asking for a meeting. You could easily do the same, or ask friends, colleagues, or direct-to-consumer customers (such as restaurants that you may sell to locally) to make a referral call on your behalf to their distributor.

Pricing to distributors can be based on an established market (such as a commodity market that you can follow; in my case it was beef and pork) or by using your own break-even analysis. Most distributors are notoriously strong negotiators, so be ready to hear a price shot back at you that you may not initially like. Distributors are experienced with margins

My advice is that you **don't let a distributor pay you more than a month out**—that is, a month after it has taken delivery of your goods. That's really asking too much of a small businessperson, I think. Also, ask up front if they'll pay upon delivery. That's a great deal better for you. If it's a point of contention for the distributor, consider offering a small discount if they will do it.

# How Rebecca Calculates Price

"A lot of people have no idea how much to charge," Rebecca says. "At first, I asked, 'How much is everyone else charging?' Then I started doing a little more research."

She explains, "Vermillion, South Dakota, is pretty rural, pretty small town, even though we have the university here. The strategies I use for pricing that I see as being effective may be different for markets in bigger cities." Certainly, she believes many of the products sold at her area farmers market are of high enough quality to fetch higher prices if they were sold in an area that could bear more premium prices.

Many farmers market vendors adopted the strategy of selling all their products below the grocery store price, but Rebecca says, "To me, that doesn't make any sense. I know how much better my product is than the grocery stores'. I check the prices at the grocery store just to see if my prices are reasonable."

Rebecca says that since the area is predominantly rural, many of the locals have farm backgrounds. A lot of older rural folks grew their own vegetables at one time or another. Now they come through the farmers market and can't understand the pricing because they think vendors are simply there to get rid of extra veggies that they can't eat or can in a season. Rebecca finds that sometimes she has to politely explain that she is doing this as a small farm business.

"It's about striking a balance between premium pricing and the customer who thinks, 'Oh, you're just selling your extras,'" she says. "I justify my prices because I know what goes into it and that it's the best there is. I know I've got a really high-quality product. Some people really understand the value, and some people come to farmers markets looking for a vegetable flea market."

and know their profit minimums—you need a little experience or someone else who is experienced to ensure that you get a fair deal when selling to a distributor.

Another thing to consider when dealing with distributors is your cash flow. When you sell retail, it is customary to be paid up front or at the point of sale. But when you sell your products to a distributor, it may not offer or even be able to accommodate an immediate payment. Expect to be offered a net-30 payment plan, where you invoice the individual in charge and receive a check 30 days after delivery of your goods.

Distributors will either pick up your product or require delivery to their location. Don't be surprised if they offer to come and get the product

She calculates price based on several factors, always keeping in mind local conditions rather than worrying too much about what big box stores are doing. Comparing with other vendors is one method; checking local retail store prices is another. She also considers how much it cost her to get the product to the market. "Sometimes I do some hard analysis, sometimes it's quite a bit softer or about how much time I put into the crop," Rebecca explains.

In the case of green beans, for example, Rebecca thinks about the cost of seed and whether she spent money on any organic pesticides, the quantity she harvests, the amount of time it takes her to harvest the beans, and the number of packaged units (bags of ready-to-sell beans) she has available. "This amounts to: What is my overall cost of producing those beans? Then I come up with a price," she explains.

Rebecca finds that she prices some things slightly higher to compensate for the other crops that just don't make enough money to cover their expenses. (There's a limit to how much people are willing to pay for each item, even if it is devilishly expensive and hard to grow. But sometimes it's worthwhile to offer a few items like this because they bring customers back.) And on certain items, she is not afraid to go a little higher if she thinks the product merits it.

"I don't have a scientific method, but I do try to think about what I should earn for my products," Rebecca says. "I also just ask myself how I feel about the price. I'm not jacking up prices just because I can, because I *know* these people," Rebecca says of her customers. "It's not like I'm sending my produce off 2,000 miles away; I have a personal investment."

themselves. Especially for food items, many prefer to pick up product and transport it themselves to their premises so they maintain control over the care and storage of the item as soon as it leaves your farm or processing location.

### Tips for Working with Distributors

- Be prepared for tough price negotiations. Find someone experienced to help you through it, if needed.
- Get referrals or introductions to distributors, or simply cold-call them on your own.

- Be prepared with a good script and pricing during your initial call. Samples are also appropriate.

- Be prepared to either deliver your product to the distributor or have them pick it up.

- Know how soon you need to be paid and ask for that plan up front. Settle for no more than 30 days out from the date they receive the product.

# Retail Pricing

Most hobby farms and local foods businesses price their products at retail because in this direct-to-consumer system, you, the producer, are the link between farm and table. Many ventures find that they are more successful, especially early on, if they stick with this pricing, marketing, and selling model. Retaining customers is also easier when you are the farm-to-table link. (More about that in chapter 10.)

Retail pricing means the price you set is the price the customer pays for an item. Theoretically, it is also the highest price paid for the item. With retail pricing, you can glean the greatest amount of profit since you don't sell to a distributor or another location that expects to be able to mark up the goods.

Many small farm ventures prefer retail simply because their production is low and they need the most money from each item in order to make the venture sustainable. As far as payment, in almost all cases, the product is either prepaid when an order is made or paid in full when product is received. I advocate this model for managing small business cash flow.

## Being Your Own Distributor

I know I just said that retail is the best way to manage cash flow in a small business, and it brings in the highest price for your product. But retail does have some disadvantages. The main one is that selling retail is a whole lot of work, and the time on task never seems to end.

When you sell at retail, you are your own distributor. There are several things to consider here. First, do you have the time not only to raise, produce, and package your items, but also to handle all the selling, marketing, delivery, and customer service? That is exactly what many small farmers are looking to do. But for others, the thought of standing at a farmers market all morning makes their knees buckle.

Rebecca's garden in springtime.

This brings to mind another consideration. Do you like people? Not just your close friends, but strangers, too? This point becomes essential to your sanity and your business success when you start to think about selling at retail. If you love the notion of conversing with customers every day and handling the good conversations and the bad, then retail will probably work for you.

Finally, being your own distributor puts a measure of liability on you that is slightly lessened when you sell to a distributor who presents the final product one or two steps down the chain. As I point out in chapter 4, you've got to have good liability insurance if the buck stops with you.

## Comparing Wholesale and Retail

There are more considerations for wholesale versus retail than I've probably thought of here, but I've covered the major advantages and disadvantages. Discuss these two pricing and selling strategies with your family and others involved in your type of business, and don't be afraid to modify your strategy once you see what is working and what is not.

The following chart has a quick list of the pros and cons for wholesale and retail.

| Retail | | Wholesale | |
| --- | --- | --- | --- |
| **Pros** | **Cons** | **Pros** | **Cons** |
| You get the highest possible price for your products. | You must devote a lot of time and energy to selling your products. | You avoid the labor and time involved in selling retail. | You get a lower price for your products—as much as 40 percent lower than retail. |
| You get paid immediately when you sell your product. | You have to deal with customer complaints and problems. | You don't have to handle customer service issues and complaints. | There's usually a lag between when you deliver goods and when you are paid for them. |
| You meet your customers face to face and can establish long-term relationships with them. | Your liability is higher. | You have access to markets farther afield than you may have been able to reach on your own. | You must work to someone else's schedule to accommodate supply and delivery needs. |
| You have complete control over how your products are presented to the customer. | Prices often stay the same (except for a sale or discount) for a season or longer. | Prices are often set daily or weekly and change as market conditions change. | The customer is buying from another vendor, and you lose a little brand identity. |
| You often have flexibility to sell when you want and at the markets you prefer to attend. | Staffing and labor become difficult as you expand. | If you use a distributor to sell your products, delivery is usually handled by someone else. | If you use a distributor to sell your products, you can't control what the final product looks like when it reaches the end consumer. |
| Your brand is always present because you are always doing the marketing as you sell. | You have to be "on" all the time with the public. | | |

## Bridging the Producer-Consumer Gap

Selling at a farmers market may not seem like a high calling, but in some ways, markets can have a crucial impact on their local communities. "Even though Vermillion is a college town, it retains its small-town atmosphere," Rebecca says. "There are university students, the town locals, and rural people. It makes things interesting at the farmers market."

She continues, "One of my biggest challenges is to try to bring all three together, the merchants, the university people and students, and the local farm people." The job of bridging the producer-consumer gap is not an easy one, but Rebecca thinks she and her fellow market vendors make a difference.

One proof is in the slow but continued growth of the Vermillion Area Farmers Market. In 2003 they started out with only one or two vendors every week. They also changed location several times and even braved some dangers when barricades were run through by motorists!

It's all part of the process, and Rebecca takes it in stride, believing in community activism. She also promotes agriculture and "civilized public discourse" on her blog at http://flyingtomato .wordpress.com and is a columnist for *Farmers Market Today* magazine.

# Combining Wholesale and Retail

Depending on what you offer and the selling venues you choose, you may benefit from combining wholesale and retail pricing in your business model, or at least offering both tiers of pricing. At Aubrey's Natural Meats, we have wholesale and retail pricing and use both every day.

Our retail price is for individual customers and restaurants that order small quantities directly from us and expect delivery or pickup at a farmers market. I use the wholesale price to sell large quantities of less saleable items to a distributor, who buys them from me at a discounted price and sells them to many locations. I also have a (slightly higher) wholesale price for larger dining establishments and restaurants that buy from me regularly and market our product under the Aubrey's Natural Meats brand on the menu.

Sometimes you can charge a retailer your retail price. Even though they are going to mark it up, they still may be able to make money on your product. This is most likely the case with high-end dining establishments or restaurants that specialize in local foods and understand your need for cash flow and competitive pricing. This strategy is challenging, because most wholesale customers will push you for a wholesale price, but there are many producers who can accomplish it. Sherwood Acres Beef, profiled in chapter 3, is one such example.

# Establishing Your Break-Even Point

Making money, or just being able to sustain your hobby without breaking the bank, is all about knowing your cost of goods sold. This is not always as simple as it sounds. While you should be keeping track of tangible costs such as seeds, plants, equipment, or salaries, there are other costs as well.

**Guard your pricing structure as strictly as you would other essential business documents and forms.** Keep any pricing you use organized in spreadsheets, business and accounting software, or simply in word processing documents. You'll need to have it handy for regular updates and changes, and to refer to what you did from year to year. It also keeps you organized so that you can evaluate your pricing track record—whether it is working for you—and make changes for each product based on your analysis.

Finally, you'll want to keep a file of any special pricing or discounts that you've offered each customer. Believe me, people will call back and say, "I want the same price as I got last year—can I get that again?"

To understand how much selling actually costs, you need to calculate your break-even point. The break-even point factors in everything, including time (especially if you are paying for someone's time), labor, supplies and other inputs, and even the delay between starting a venture and actually making money. For example, fruit trees and wine grapes take several years to yield a return, while corn and tomatoes produce in a matter of months. The same thing goes for livestock: grass-fed beef or bison take several months, but not perhaps as long as acquiring nanny goats, raising them, milking them, and then crafting aged cheese from their milk.

One strategy for calculating the break-even point is to add up all your costs and then average them out across the number of items you'll sell. In my case, I added up the total costs and divided that by the number of beef and hogs I was moving per month or per week. This gave me a cost per head to work with, and I could begin considering whether to increase price or cut costs from there. The equation looks like this:

Total costs ÷ Total units sold = Break-even point

## Honest Pricing Practices

When it comes to direct-to-consumer selling, especially at small venues where there is competition among your peers every week, there is an unstated etiquette to observe about pricing.

Rebecca thinks vendors should do what they want to about price, based on their products' uniqueness and quality and on how well established they are with customers. "I have enough of a reputation now that I can charge what I need to charge and not worry about what other vendors are charging," she says. "I may lose some sales that way, but I don't lose many."

Pricing relative to quality is one thing. But dropping the price just below the competition to snag sales—that is to say, price gouging—isn't a practice Rebecca is fond of. She also believes vendors shouldn't get into the habit of allowing customers to nickel and dime them or pit one vendor against another.

Another problem that arises from time to time is vendors marketing their products under a specific label, such as USDA Certified Organic, even though their fellow vendors know that they are being, shall we say, a little less than honest. Because Rebecca is not certified organic but uses many of the practices organic farms do, she uses an unregulated term for her crops—sustainably grown. She then discloses as many details about her production practices as the customers would like, even inviting them for a farm tour if they're interested.

Price gouging and being dishonest about a product's provenance can be problems at any farmers market, and they can really affect the bottom line of all the merchants there.

Being honest with customers and respectful of fellow vendors is all part of pricing and selling products, according to Rebecca. As the president of the Vermillion Area Farmers Market for the past six years, she deals with issues as they come up, but wants all vendors and customers to feel welcome and able to have their own strategies, too. "We're trying to encourage a culture of positivity," she says diplomatically.

## Factors to Include in Your Break-Even Analysis

- Time for everything involved in production, sales, and service
- Specific costs for all raw materials
- Taxes, interest, and payments on land, equipment, and other business items

- Expenses involved in getting the product to market and getting it sold, including raw materials, fuel, electricity, marketing materials, and advertising

# Pricing Strategies

The first step in creating a pricing strategy is to give your price a base. You need to decide: What is my price based upon? This base rests on your break-even analysis—knowing the cost of goods sold. The next step is to do a little comparison shopping. Go back to your market assessment and give some thought to demographics. What are comparative products selling for in your area, and what are the people to whom you'd like to sell willing to pay?

**One of the best ways to improve your break-even point** is not to raise prices but to reduce expenses.

If you'll be offering wholesale prices, visit commodity markets and look at the pricing for similar products. With Aubrey's Natural Meats, there were already a few established national brands of natural beef and pork. I looked at their pricing to get an idea of what distributors were willing to pay, then decided how much more valuable I thought my features and benefits were. Browsing online is easy and useful market research. Ultimately, you are looking to compare prices to help establish your pricing base, not to copy what others are doing.

The next step in pricing strategy is to insert your own intrinsic value into the base you've decided is workable for your market. In seminars, this point always brings up the questions: What is the value of artisan? How much is hand-crafted worth? How much more is my quality, effort, or benefit worth? The short answer is, whatever someone is willing to pay.

The long, vague answer is that it depends on your situation, the uniqueness of your product in the marketplace, your ability to market and sell, the demographics of your customers and their willingness to pay, and even the area where you live. I cannot answer for you how much your specialty is worth, but I can tell you that you need to factor it in, mark your product up accordingly (whether that markup is 10 percent or 100 percent), and take that markup as your profit.

## How Much Pricing Information Should You Publish?

Is full disclosure the right way to go when offering pricing information to the public? The answer may seem like an easy yes, but I'll challenge you

What could be more satisfying than to be surrounded by the literal fruits of your labor, as Rebecca is here?

just a bit on this. You have some things to consider when deciding how much of your pricing information to make accessible all the time and how much to disclose "upon request."

If your products are available only at the market, for example, or are not available by pre-order, then you may not need to have price sheets available online, via email, or in printed form for people to take away. You'll probably just post your prices at the market. This setup gives you flexibility to change prices often without having to update numerous places where they are displayed or posted.

If you offer online ordering and payment, you'll probably need to keep all your pricing available so you can smoothly conduct transactions. The exception is when your business is seasonal. In that case, it's wise to make pricing accessible only for the season in which you do business and take the prices off the web site after your selling season ends. This way, you avoid haggling with customers who say, "It says this price on the web, so that's the price I want!" You need to be able to recalculate prices from season to season, and taking the old prices down makes doing so a lot easier.

# Establishing a Web Presence

I asked Rebecca about getting started with blogging as it relates to starting a small business. This is what she wrote:

Establish a web presence and get a digital camera capable of taking high-resolution images. You don't need to be a web designer or hire one. There are lots of good free weblog ("blog") platforms that you can set up in a matter of minutes. I use Wordpress.

A presence on the web enables interested people, customers, and the media to find you. You can also link to other blogs or web sites that are doing similar things to establish a network, meet other producers, and get more traffic to your site. You can use a blog as a journal of farm happenings and projects both for yourself and your customers. Good blog platforms enable you to restrict personal information you don't want advertised and keep posts private in more of a diary-like format. They also help block spam and allow you to moderate comments to block any objectionable content before it shows up on your site.

You can use the blog to let people know what markets you'll be attending and what you'll have for sale. I have had a number of new customers at farmers market who stumbled on my blog and then showed up at the market to buy directly from me. It also gives your CSA members and market customers a feeling of connection to you, your product, and your farm, and it gives new customers enough information about your operation that they feel comfortable approaching you and asking questions.

Along with farm project updates, I post each week's CSA newsletter on my blog. That way, potential new CSA members can see what each week's delivery contains and what kinds of recipes and tips I include with each delivery. Current CSA members have a backup in case they lose their hard copy and want to retrieve a recipe or a storage tip from earlier in the season.

The digital camera is great for visually recording projects step by step, recording successes (and failures), and also to be able to consult experts about problems you might be having with insects or disease. It's also good for being able to supply images to the media in case they want to do a (free publicity!) story on your operation. You can also easily upload images to your blog or web site to add visual appeal to your products and help potential customers to recognize you on sight.

CSA bags are loaded and ready for delivery. "I always had the idea of starting a CSA in the back of my head since I worked on the first one," Rebecca says.

Finally, if you are planning to accept only custom orders, whether online, in person, or at farmers markets, it probably makes sense to limit what you publish about your prices, since you may want to price your items by the order. In this case, offering a range of prices is best to give interested people an idea of what they might be spending without committing yourself.

## Sales and Discounts

Once you know your break-even point, you can always decide to move up or down from the price you set based on the profit you'd like to make in relation to your loss potential. Knowing your break-even point is especially important when you consider offering sales or discounts. Compared to the products you buy at typical retail stores, there is less discounting in the direct-to-consumer market. That doesn't mean offering a sale isn't a good idea; it really can work to move products while still protecting your bottom line.

## Rebecca's Tips for Changing Prices

Coming up with a starting price can seem difficult enough, but your next job is to keep changing with market conditions and your own need to make money. Rebecca agrees that changing prices can seem challenging, but it's a must to consider, at least after the first season.

"I try to keep my prices the same throughout a single season—unless it was like this year and I went to the grocery store and thought, 'Oh my God! I'm not getting enough for that!'" She was asking just a quarter apiece when her local store was asking $1 to $2 for cucumbers that weren't nearly as fresh.

"I did raise prices based on what I saw at the grocery store," she admits. But she did so moderately, coming down from the store's price just a bit to fit what she thought her market would bear.

Rebecca does look at prices at least at the beginning and end of every year, considering what her expenses will be and what changes will cost her extra to get crops to her CSA members and farmers market customers.

Most of the time, Rebecca's customers have been very receptive to changes in price and understanding of prices going up. She feels that a good way to handle any negative customer reactions or confusion is to be up front about the changes and the reasons for them. "I'll explain why there's a price increase to customers. Then I'll say something humble like, 'This is what I feel I have to charge.' They'll often buy from you because you took the time to explain."

There are some common ways to put farm items on sale. One possibility is packaging several related products together, such as in a combo pack of items needed to prepare a certain recipe. Another idea is partnering with another vendor for a complete package that includes all items at the best price. This would work especially well if you provided recipes for an entire meal and customers had to purchase an item or two from each of several vendors to prepare a balanced meal of protein, fruit, herbs, vegetables, breads, sauces, and even table decor, such as a floral arrangement.

Bulk and quantity discounts are common in small farm business. With Aubrey's Natural Meats, some of our best sellers ever were bulk ground beef and ground pork or sausage. I offered 40 percent off the retail price for orders of 25 pounds or more of those items. This program was wildly popular and helped me move products I had in higher quantities.

Another discount strategy is to offer a better price if the customer either prepays or pays in full by a certain date. Again, much of small farm business is managing cash flow, so if this strategy generates sales, makes money, and keeps cash in the bank, give it a try.

## Raising Prices

When you go to the store, you're usually not surprised if the prices of your favorite items go up from one season to the next, right? Shouldn't it be the same for small farm businesses?

In theory, yes. But for some reason, convincing our customers of this seems to be more difficult than it is in "regular" business. People expect the same item to cost the same price, year after year.

I've met with resistance when I've tried to eke out another dime per pound. I'm not exactly sure why many consumers resist periodic price increases, but they do.

The irony is that producers may also be resistant to price increases. One theory I have is that the relationship between producer and consumer becomes so much closer when you know your customers that many producers feel uncomfortable asking for more money. Within that personal relationship, asking for more can be tough. Plus, small farmers are already grateful for the business.

That said, you can and should change prices if you find that they are not working. And you can raise prices when you need to. Again, the key here is knowing your break-even point and your profit goals. You'll also need a bit of finesse to deal with your customers. One strategy is to raise prices on a set schedule and to publish that schedule as something like "2010 Summer Prices, 2010 Fall/Winter Prices, 2011 Spring Prices," and so on. When people see this, they can understand the differences based either on seasonality or on periodic raises. People like good surprises and get uncomfortable when they get surprises that cost them money—even if it's only a dime. If they know the increase is coming, it's not so bad.

## Rebecca's Best Practices

- Always maintain some measure of control over the land you're using.
- If you don't own the land, be sure you have a written agreement for the use of the property.
- Set prices based on what your market will bear.
- Research prices locally at the grocery store and by observing what other vendors are doing with their prices.
- Don't be afraid to change your prices if you need to, even midseason. If you do change prices, volunteer an explanation to customers about how your product merits the increase.
- Be honest about your production methods. If you're not USDA Certified Organic, for example, don't say that you are. Instead, offer another explanation of your production practices, such as sustainably grown or all natural.

# Chapter 8

# Where and How to Sell

**Learning Objectives**

- Discover a variety of selling venues that are especially useful for hobby farms and local foods businesses.
- Learn tips for using these venues successfully.
- Understand how to combine several selling venues to maximize customer contact.
- Learn how to manage labor.
- Uncover creative internship and apprenticeship ideas for young people on the farm.
- Meet Denise Baarda and learn how she operates her Veggie Van, creatively serving customers with special needs.

## Sales and Marketing Intertwine

In the next two chapters I'll review the twin towers of being connected to the customer and to commerce—selling and marketing. These two activities work together to get you customers, get you paid, and keep customers coming back.

Sales and marketing are two different things, but you will find them blending in your business all the time. For example, the act of selling is actually a form of marketing, since you are in front of the customer during the transaction creating an impression and a perception of your products and your company. Marketing is also a form of selling, or, at the very least, is part of the sales process. When you're marketing products and services, you are doing so with the intention of getting something sold as a direct result of your marketing activities.

Meat and cheese are obvious add-ons for the shop at this California winery.

In this chapter I'll look at selling venues and issues surrounding selling direct to the consumer from a small hobby farm. Then, in chapter 9, I'll cover marketing ideas to secure your success at the locations you've chosen.

## Places to Sell

So where do you want to sell your stuff? Believe me, the venues for local foods businesses and small farms are growing in both number and variety. In addition to some established venues that I will discuss here, you are limited only by your own creativity when it comes to finding the right place to sell your home-raised goods.

One of the great joys of working for yourself, especially in rural business, is that the "conventional" commerce system need not be your playing field. There's no need to sell by standing on your feet all day at a retail store if you don't want to, nor do you have to allow people out to the farm if you seriously value your privacy. Whether you think shivering or sweating at a farmers market early in the morning is pleasant or painful, you choose to use that selling strategy or come up with your own. The model you choose reflects the style you bring to food and farm, your values, and your personality. Being new to farm living means your wealth of new, nonagrarian perspectives is awaiting interest from your customers.

As I've said before (and it's so true), being a consumer yourself before you became a farmer or a foodie is a major advantage when you're conjuring up unique ideas for taking a product to market. Let that uniqueness shine through; it is an asset.

## Finding Selling Venues

How do you select a selling venue? Where do you find ideas? Every area is different in terms of what it offers, of course. I always advocate carving out your own niche if something you want to do isn't already available. Until then, to investigate the current marketplace you need to know where to look.

Start by contacting local trade associations, rural development groups, and local "main street" or community promotion organizations in the towns

Denise Baarda (far right) offers a variety of products to keep customers interested. She grows some herself, and others she buys to resell. The Veggie Van is in the background.

## Meet Denise Baarda, Small Farmer

Denise Baarda of Mount Bethel, Pennsylvania, comes from a long line of vegetable and fruit growers, but she herself didn't start out as what one would call a farmer. Denise has always been drawn to plants, though, and she has an associate degree in florist management and business. In 1984, at the young age of 16, Denise opened Plaza Florist in Newton, New Jersey, with backing from her parents, Marianne and Nathan Gould. "I really had to prove myself to the whole area that a 'kid' could succeed in business with proper backing," Denise recalls.

Apparently, her farming heritage was so strong that working with floral arrangements was not quite enough. "We're no strangers to the growing field," says Denise of her family. The Goulds are actually the fifth generation in the area; her great-great-grandparents arrived in the area with "a handful of beefsteak tomato seeds, bell pepper and basil seeds, and a variety of flower seeds, all to make a start in the new land," she explains.

Denise's ancestors worked hard to become growers, eventually supplying upscale florist shops in New York State. Her relatives also believed in a spirit of sharing, and during the Great Depression they kept greenhouses for themselves and others. "They fed those who they knew truly needed an extra hand. I'm very proud of my family's growing practices. And they never did any deals without a hand-shake—a trust that in today's market is rare," Denise says.

Eventually, Denise and her husband, John, decided they wanted their own farm. So they bought an old, dilapidated orchard that Denise says locals called the River Orchard of the Shoemakers, referring to the family who once lived there. They opened a produce shop to make ends meet while the orchard came back to life. "When we moved back to the farm to get it in working condition, we sold the florist shop. But, to my amazement, I found that I had a

or cities near where you want to sell. These types of groups are the largest sources of information on festivals, fairs, farmers markets, and other selling venues that unite producers and consumers at an established location or activity. Find them by going to those activities and events yourself, calling peers, and searching online.

Next, think about what kinds of goods and services you like to get when you go shopping, and where you like to get them. What do you think is missing as a distribution venue in your area? Consider options such as home or office delivery, online mail order or online ordering with a pickup following, so we included a florist in our produce store," Denise recalls.

For seven years, Denise and John and, eventually, their children, Jacob and Sabrina, worked to get the old place back into production. Denise says they focused on sound agricultural techniques that helped them manage their lands sustainably. As a former florist, she had to get used to the idea that fruit that wasn't perfectly pretty was still good eating.

"Now we grow all types of berries, along with all types of fruit trees, seasonal veggies, and winter squashes," Denise says. "But the best addition and customer attraction that we added is a cider house. At the farm we press our own cider." The day I spoke with Denise about this book, she was pruning fruit trees in the blustery winds of late December while her dad mentioned that adding another 100 trees would be a good idea.

Cider has been a big growth area for Denise and John, and she is very proud of their product. "The best part is the cider isn't pasteurized. So many people travel to our farm for that reason alone," Denise says. She is quick to explain that the cider is still safe to drink. "It gets inspected and water samples are done, we have 'serve safe' classes, we use stainless steel, and so on and so on." The cider is popular with locals and people from out of state. "To have customers travel from New York City just for your cider gives you a wonderful feeling," she says.

Now the family farm includes an orchard set on 20 acres in rural Northampton County. The family eventually expanded the greenhouses, hothouses, and field crops to branch into local farmers markets and flea markets. The florist shop stayed as well. "Now we have a florist, gift, and produce farm market right here on the farm that is open year-round, and the public loves it," Denise says proudly.

location or local delivery, shopping in person at a retail store or market, shopping for convenience with one-stop sources, and shopping for pleasure such as outdoors or at a specialty venue. Knowing the types of items you plan to sell, which of them fits with your sales and service ideas, and why? Are other, similar items selling well at a given location? If so, why is that?

Don't overlook the softer side of selling; consider some of the major tourist attractions in your area. There may be a selling venue for you at a tourist stop or visitor center—maybe a nice retail boutique or a Saturday morning market in the summer. Capitalize on regional fairs and festivals that celebrate everything from local heritage to the annual harvest. If nothing looks good, think about starting your own event with a few neighbors or other like-minded, noncompeting small farmers.

**One great online research stop is the chamber of commerce in your community.** My experience is that chamber of commerce staff are always delighted with any new business, no matter how small. They'll welcome the chance to introduce you to their community and assist with all kinds of business-building activities.

To search for these types of venues, contact your county or state's tourism board and ask them if they've worked before with a business like yours. If not, ask them to meet with you to discuss your ideas. Tourism boards are always in the market for new products, services, and willing people to further their mission and keep products and activities for the region fresh and inviting to visitors.

In this section you'll find a variety of selling venues you may want to consider for your products. Each one includes both the obvious advantages and the potential disadvantages that I've seen from experience.

Baarda Farms gives seniors a chance to independently select produce and meet with the farmer directly, all without having to drive to a farmers market.

# Farmers Markets

The most common direct-to-consumer selling venue today is the farmers market. More than likely, you've been to a farmers market, and you're probably a fan. As I explained in chapter 1, the growth in both the number of farmers markets and the volume in sales has increased steadily over the past decade, due greatly to the fact that consumers love to meet the producers and buy their foods fresh and direct.

Obviously, the idea of selling at a farmers market is not innovative, but it works. Participating in a farmers market often doesn't cost much (the fee may be as low as $10 per day), and it offers a great way to make money and advertise your foods or goods at the same time. Farmers markets combine sales and marketing every day you're there.

Farmers market setup is quite simple and can be extremely inexpensive. A table, a scale, and a cash box are the basics. You can spend a bit of money on banners and signs, and even have a logo designed to showcase your business name and make it memorable. You'll also need some protection from the elements; many markets are held outside even in the rain. And be sure to contact the health department of the county in which your market is located to get a temporary permit.

There are drawbacks to farmers markets, including labor time and the perishability of foods. You also have to decide who will be working at the farm while you're selling at the market. (More on labor later in this chapter.) Plus, the market is only as good as the amount of traffic it gets. Other drawbacks include the fact that you have to pack, unpack, and then pack up again, every time you attend the market. And it can get cold, hot, buggy, humid, windy, rainy, and even snowy when you sell outdoors.

## Types of Farmers Markets

Farmers markets come in a variety of styles to fit customers' preferences and vendors' production seasons. I'll list some common types. Most producers I know combine several types, if possible, and attend more than one market a week. At Aubrey's Natural Meats, for example, we routinely attended up to five markets a week in the summer, so customers can see us at the most convenient location for them, on the most convenient day and time.

### Seasonal Markets

Seasonal markets start around the first of May (or whenever flowers and early produce such as lettuce are available) and continue through pumpkin and apple season, or sometime around mid-October. A newer trend is the winter market; I've seen successful winter markets even in places where winter is, well, quite wintry. These loyal vendors and customers will stand

## The Veggie Van Hits the Road

"Our newest idea in farming is the Veggie Van, a mobile farm market that caters to our local seniors," Denise says proudly. "The whole idea of the Veggie Van came from customers who came to the store at the farm, then asked us to drop off a few things here and there for them in town." More than just a delivery vehicle, the Veggie Van is a mobile store on wheels.

"It's a utility cab that we just put on the back of our pickup truck," Denise explains. "When we get to our destination, my daughter, Sabrina, and I unload tables and set them up as if we were at an actual farm market." Their primary market is seniors, and they mostly take the Veggie Van to assisted living communities.

Denise leaves the packing and setup to Sabrina. "My daughter is *beyond* organization. She puts everything in alphabetical order, including our fruits and vegetables. We have everything in the pickup truck in either picking baskets or apple crates. The biggest trial and error we face, not only with the Veggie Van but also on the farm, is how much product to bring and how much to pick so there isn't too much waste," Denise says. "Our routine is to pick fresh daily, early enough so if we run out, we can run to a field and get more!"

So far, the van has been pretty successful in serving customers who otherwise might not be able to get out and shop. While this is a great service to residents in assisted living communities, getting permission to come wasn't always easy. Luckily, Denise has plenty of

outside with greenhouse-grown cool-season veggies and other all-season items such as eggs, cheeses, and meats.

Seasonal markets are great for new entrepreneurs and those who get busy at the farm or work other jobs, because they usually require a commitment for only a few months. For example, a location may host a summer farmers market, then take a few weeks off and begin a winter-spring market. You can choose to participate in just one or both.

### Year-Round Markets

Some markets are held year-round. Usually they shut down for a week or two around the major December holidays. These markets are either held inside in a heated venue that offers more amenities, such as bathrooms, or in areas of the country where it's warm and sunny all year.

experience in getting her products placed. The Veggie Van is a product of her persistence.

"A lot of my sales pitches were physically going and introducing myself and giving a whole spread on how and why the idea would benefit those who can't drive or walk to the local stores. I used a couple of poster boards with photos from the farm and pictures of us actually working and picking," Denise says. She adds that care-givers appreciated the fact that Denise and her daughter, who are the farmers, would also be staffing the van.

Denise had to get approval from her state and from local hous-ing authorities for the Veggie Van so she could sell produce to sub-sidized housing facilities for seniors. She accepts nutrition vouchers provided to seniors by the state's Farmers Market Nutrition pro-gram. Denise also arranged to have the Baarda farm inspected by the state department of agriculture; with that approval many of the senior facilities felt comfortable about the Veggie Van stopping by.

Denise views the Veggie Van as a service, and as part of that service, the Baardas host many seniors at the farm each year. "We invite our seniors from the various homes to come up and tour the farm," Denise says. "We built an arbor with tables to give them a beautiful view while they eat, since they bring lunches or coffee and cakes. To see and hear the comments about how the overlook of the Delaware River is so breathtaking makes us happy."

The major difference between a year-round market and a seasonal mar-ket is the expected commitment. Year-round markets usually require long-term contracts. They may also cost more because of the better facilities. They may even require that you join a market association or some other membership group.

### Commuter Markets

These markets are often held on weekdays in the early evening, to attract people on their way home from work. If you already sell at a Saturday morning market, a commuter market can be a great add-on. You'll likely find these markets to be family oriented and to feature "just in time for din-ner" marketing.

Commuter markets are a good way to find new customers who are too busy on Saturday mornings traveling around to their kids' activities or

## Trying on Selling Venues

Over the years, Baarda Farms has worked with a variety of selling venues. The primary one is the retail store on the farm. It's open year-round, seven days week, from 9 a.m. to 5 p.m. They offer vegetables and fruit, plus eggs from more than 300 free-range chickens.

They have delved into agritourism and find that most of their ideas are successful. "We offer nursery school tours, which are held in the spring for planting season, and then I also offer a fall harvest tour that includes how we make our cider and how we pick apples," Denise says. They tried doing a U-pick farm, but Denise found she didn't like the way some people failed to respect the farm and their personal property, so they decided not to do it anymore.

Farmers markets and flea markets in the region are their main offsite venues. But as the farm has grown, so have the ideas from Denise and her family.

They try to select venues based on what they know about the location. But sometimes knowing whether a selling venue will work out comes from trial and error, or at least after a season or two of experience. A selling venue that makes the customer *and* the seller happy—now that's a good fit!

**Commuter markets are increasingly popular in urban and suburban areas,** where they are held near public transportation hubs. You'll find that the shoppers there are particularly grateful for the access to farm-fresh products.

value sleeping in too much to get up and go to the market. You'll reach different demographics, too, which also generates sales. For example, if you attend a commuter market in a church parking lot, you'll not only see locals and those driving by on their way home from work, but also benefit from the business of church members who are there to support their own function.

Most commuter markets are located at key travel intersections, and I find it's best to attend these types of markets in larger, more populous regions where people gather at the end of the business day. Unless you've been to a particular market and seen great traffic and brisk sales, I'd steer clear of weekday markets in small towns or areas of the city where people don't gather, shop, or party after work. I once sold at a market like this where the venue was quaint and the people running the market were nice, but it was in the somewhat beleaguered downtown of a small city and the traffic was really, really poor. Even though I wanted to support the venue, sluggish sales and a feeling of time wasted made me pull my products after just one season.

If you work another job, of course, the commuter market may not fit; common times are from 4 p.m. to 7 p.m. on a Wednesday, Thursday, or Friday afternoon.

### City Markets

A city market is typically a large retail venue in which vendors formally lease space in the market that they keep stocked and staffed every day the market is open. This type of market has been popular on the East and West Coasts and in ethnic areas of cities for two centuries. Now, a growing number of midsize cities in America's interior are either revitalizing old, formerly defunct city markets or refurbishing old buildings and creating new selling venues.

The city market is commonly located downtown within walking distance for many downtown workers and residents. City markets don't sell only farm-raised products, either. Many are in lovely, historic buildings that offer everything from office space to art galleries, restaurants, and gift shops.

While I really like the idea of a city market, I'd recommend visiting the one you're considering a few times first. The trend is catching on, but some folks I've spoken to report slower sales as compared to their established Saturday morning farmers market. Also, consider the type of product you sell. Items that are large, heavy, or perishable may be a challenge to sell in an area where the customer is walking several blocks to and from the market.

### Local-Only Markets

Here's a niche that didn't used to be a niche—which, of course, is the essence of the entire small-farm movement. Whether it's because artisanal vendors are fed up with chintzy knickknacks clogging up farmers market stalls (making them appear more like flea markets) or because customers are demanding it, many markets are now either started on the principle of locally grown and made products only or are implementing new rules that require verification of the product's origin.

Personally, I love this concept. And with your new hobby, it will probably serve you very well to combine selling and marketing on your "localness."

Sabrina (right) and Denise (next to her) are ready for senior shoppers.

Just how local-only the market is seems to depend more on the availability of a nice selection of goods than the strictness of the market masters. For example, in states such as Indiana that have some climatic differences from north to south, many local-only markets specify "grown in Indiana" rather than using a mile radius or a county distinction. Most customers appreciate this as a way to support small farmers while still being able to get southern Indiana-grown melons in northern Gary, Indiana.

## Who Shops at Farmers Markets?

Speaking of local, that is typically the top reason shoppers turn out in many areas year-round to buy at farmers markets. We all know that the local foods movement is alive and growing. But farmers markets are a special component of that movement: They complete the circle from farm to table. And on those warm summer days or brisk early mornings, the urban-rural gap is narrowed as farmers markets provide a regular location where the intertwined farmer-consumer community thrives.

We've been working farmers markets with our business, Aubrey's Natural Meats, since 2004, and I can tell you firsthand that the variety of customers who choose farmers markets is more varied than any group of people you'll see at a state fair. Our loyal customers range from soccer moms to CEOs, from chefs looking for fresh ingredients to use that night to large families and church groups looking to stock up on a half or whole side of beef.

Farmers markets are accessible to all levels of people. And with the uniqueness of products and competitive pricing that many vendors offer, they appeal to people who are seeking a rare find as well as families on a tight food budget.

It's true that customers can find "gourmet" food items at a local market, but lately I've read a number of articles pointing out that vendors at farmers markets often charge less than their big-store counterparts or even local grocery chains. This doesn't necessarily mean low return for the vendor; according to the USDA's Agricultural Marketing Statistics Service, producers selling at farmers markets nationwide bring in a collective $1 billion annually!

## Farm Stores

Farm stores have a lot in common with agritourism and are certainly an agritourism-type idea. In fact, many times the store is combined with a tour or a U-pick opportunity. But a store is still a selling venue. Tourism aside, many farms are creating quaint and convenient (for them) shops right on the farm in old buildings ranging from well pump houses to barns to grain bins. Your remodeling skills and the zoning requirements you must follow are the only limits to how you set up a store on the farm.

Successful on-farm stores should post hours of operation and keep yourself or a staff member handy during those hours. If you don't want to have someone sitting in the store all the time, you'll have to set up a device that rings or beeps where you are to let you know that you have a customer.

I know one woman who offers her ground beef on the honor system. She leaves the door and the freezer unlocked, asks the buyer to jot down how many packages they bought, and leaves a money box for checks or cash right there. Once a week she hangs around the store to greet customers, but the rest of the time, they come and go as they please. So far, it has worked for her.

One big caveat: Watch the liability of having folks on the farm and protect your assets.

**If you have a store on the farm, make sure you do a lot more marketing offsite.** When you sell only on your quiet country lane, you won't get the built-in marketing you get just for being at a farmers market. I recommend spending some money on targeted advertising and starting with conservative sales goals.

## Retail Stores

Some people dream of being a shopkeeper, and other people think being tied down to a store every day is like the old farmer's wife's saying, "He's married all right; married to the dairy!" The axiom is true: Farm living gives you less freedom to leave the farm because there is so much to manage and take care of. It's the same with a retail store.

Still, for many food ventures, opening a retail location or putting your products into a successful shop is a great way to boost sales. One good reason to sell in a retail shop is that it's different from farmers markets, so you'll reach a different demographic and add customers. The store also offers conveniences that markets don't, including year-round storage and on-site supplies. Not having to pack up to go to the market, pack up to leave the market, and unpack when you return home is appealing.

The big consideration with a retail store is the price of the lease. How do you know how much you can afford to pay if you haven't sold anything yet? The short answer is, you don't. This particular fact is one of the details I overlooked when I set up my store. It was just too soon after I started the business to know how much I could afford to pay in rent and how much money I could make. For that reason, experience tells me to recommend that you give your new venture time before spending a great deal of money on a retail shop.

## Selling on Consignment

A cheaper alternative to opening your own retail outlet is consignment selling at someone else's store. When selling on consignment, you take your product to an established retail store and leave it there, and their employees handle the sales. This is not the same as simply convincing a retail store to buy and stock your items. Consignment selling means you set up your products in a space they provide. You set the price, too.

You will receive a check weekly, monthly, or by some other regular arrangement you negotiate. Of course, you'll pay the store a fee for selling the product for you and allowing you to use their space. Consignment fees can be from 10 percent to 30 or 40 percent. So, for example, if you sell goods priced at $100 in the store and their fee is 10 percent, you would get $90.

Common consignment fees are 15 to 20 percent, and that seems fair. Personally, I don't think they should be over 30 percent—unless you have a really big profit margin that you're willing to cut into. Any more than that and the store is taking more than it's worth, in my opinion.

While they'll handle the transactions and may do some restocking for you, don't rely on the employees or the store owner to do as good a job as you would. Stop by and call regularly to see how the display looks and how your sales are coming along. Be organized and accurate with your initial and ongoing inventory management. And make sure they have a reliable system in place to track your sales.

As far as setup, sometimes you provide everything from the display case and refrigerator or freezer to the bags to package your product, and sometimes you just put the product on the store's shelves or into the coolers. Either way, make sure your signage is obvious and different from other signs in the store, and is displayed prominently. Insist that you are allowed as much signage as makes sense to get people to the case to pick up your product and recognize it as yours. Be firm on this point. I'd consider not selling in a store that won't allow you to market yourself when they are not buying the product outright.

Commonly, stores will arrange to offer your product for a few months to see how it goes, returning it to you if they feel they can't sell it. Likewise, you should be able to pull products from the shelves at any time.

This model is gaining in popularity in the wine industry. Tasting rooms are very expensive to set up and require permits and licensing that are often too expensive for limited-production wineries. So they take their wines to a retail venue tasting room that offers several small brands and handles tasting, shipping, and sales. I believe this concept could work with all manner of farm-raised goods.

## Home Delivery

Does anyone reading this book actually remember having a milkman or a paper boy? I have to admit, I'm not old enough to recall either. Those conveniences were once commonplace, but nowadays we associate them with *The Andy Griffith Show* and the Mayberry lifestyle. Well, here we are once again, and what was once old is now new—or shall we say *nouveau*—in the world of local foods and hobby farms.

Home delivery of many foods is gaining acceptance in a hurry. Of course, take-out food like pizza has always been available for home delivery. But now fresh foods are making their way to doorsteps across America. Whether you raise the products yourself or source them from your farming neighbors and mark them up, this type of service works for many entrepreneurs. It's a perfect food and farm business for an entrepreneur who prefers to remain in town and wants to be a distributor, not a producer.

Delivery businesses are as varied as the people who own them. The simplest type of delivery is akin to what I offer through Aubrey's Natural Meats. When customers order a large quantity, such as a half beef, it's difficult to take that much meat to a local farmers market for pickup, so we offer delivery for a small fee. We also offer delivery on orders over a certain dollar amount in the winter, when our farmers markets are closed.

## Working with Seniors

"The original Veggie Van, serving our seniors" is the slogan painted on the side of Denise's truck, along with brightly colored logos of a chicken and produce. The van gets noticed and brings welcome goods, much like an ice cream truck on a hot summer afternoon.

Senior living managers have told Denise that the Veggie Van fondly reminds many residents of the days when milk was delivered to their homes. While some residents don't get out as often as they used to, knowing the Veggie Van is on its way has encouraged many and given them something to look forward to.

Denise caters to the special concerns of seniors in another way, too. "Basically, many of the seniors are on a fixed income, so we price according to the area we are visiting that day," she says. Denise also adjusts the container size so there's enough for just one or two meals, because many seniors are cooking for one. "They buy $5 to $8 worth of food to get them through until the next week, when they know we'll back with fresh produce."

Delivery can be direct to the customer's home or office or to a delivery point such as the nearest farmers market. Just make sure to contact the owner of the parking lot where you'll be meeting customers and get permission. Most places won't charge you, especially if you are courteous enough to make arrangements in advance.

A newer type of delivery service provides a week's worth of groceries of all kinds, sourced from a variety of farms and vendors. This kind of food delivery is increasingly popular with urban and suburban Generation Xers who are in their prime working and family years and put a premium on quality, wholesomeness, and, especially, time. In urban areas, for example, "supermarkets" without retail space are taking orders over the Internet and then sending trucks to deliver the goods. A web search for food delivery in your local area will undoubtedly yield results that will give you lots of ideas for delivery services you can sell—or confirm that your area is in need of this kind of service.

Delivery offers convenience, and convenience means repeat sales, which is why this selling venue is so popular. One of the main drawbacks is labor. Someone has to take the orders, package them, make the deliveries, and either collect payments upon delivery or invoice the buyers. This sounds easy, but invariably customers will call and cancel when you're already on your way, or may just not be home. These glitches can be maddening. Always call before you leave the farm on the morning of the delivery. People can use a gentle reminder, and you'll value the time saved.

The other main drawback is storage space, especially if your product is seasonal and can be stored for a few months, such as apples or wine, or takes up considerable refrigerator or freezer space and uses energy, such as meats and cheeses. Plan to have enough space and proper storage facilities to hold products in inventory that are awaiting delivery.

## Internet and Phone Sales

Move over, eBay—hobby farms are here! Now that just about everyone has access to the Internet and buying online is widely accepted, offering some or all of your products online and then shipping them seems ideal. Since some people still don't like Internet shopping, or feel uncomfortable putting their credit card number into cyberspace, offer a phone number so these customers can call to place their order.

There are several benefits to offering Internet sales. Among them is the fact that customers can order whenever they have the time, even if it's in the middle of the night. You can also expand your customer base—in some cases, nationwide. Think about how many boxes of oranges are shipped from small family groves in Florida to just about every state each winter.

I love these kinds of orders because they help so much with cash flow in a small farm business. Knowing what must be sold and having it paid for in advance is a nice safety net when you are starting out or at the beginning of a season. Orders also provide excellent tracking so you can look back and see just what sold and whether any trends are associated with those sales.

Before you decide to sell and ship online, find out whether you can ship your product across state lines. Things like meat and dairy that are produced only in state-inspected (not federally inspected) plants cannot be shipped out of state, for example. And there are many, many regulations concerning wine shipments. Check with your state's board of health and revenue service first!

With ordering comes shipping. Of course, you could do as I do with Aubrey's Natural Meats—take prepaid orders but not offer shipping (I deliver within a relatively local area). I also accept orders over the phone, and when I get the customer's check, I process the order. But for Internet orders paid with a credit card and delivered anywhere, you'll have shipping costs.

Work with a reputable vendor to get a good idea of the costs associated with shipping, the packaging required, and box and supply costs. Decide whether you'll take the orders to a shipping center or buy the shipping materials, package everything yourself, and get regular pickups. Shop around for these services, because prices and rules vary. It's smart to pass along most or all of the cost savings to the customer. If your shipping is too expensive or too inconvenient, you'll get only one order from that customer—I guarantee it.

> If regulations or practicalities mean that you must limit your shipping to certain areas, make sure you spell out those limits very clearly on your web site. **State exactly where you will and will not ship.**

Other management decisions you'll have to make include how to insure the packages, whether to pass on that cost to the customer or absorb it yourself, and how to deal with spoilage. The cost associated with credit card use is another issue that comes up more often with Internet orders than if you sell at a market or take only cash and checks at your on-farm store. Finally, returns and customer satisfaction may become a larger issue when you're shipping products than when you can see for yourself the condition an item is in when the customer receives it. (I talk more about returns and exchanges in chapter 10.)

Shipping concerns aside, I am convinced that the major disadvantage of this selling model is a lack of customer contact. This is a serious consideration, because I find that direct, regular customer contact is what breeds loyalty and continued enthusiasm for your products. One way to allay this problem is with newsletters and regular Internet correspondence. (I cover customer retention and communications in detail in chapter 10.) But there is no substitute for getting to know your customers personally. If knowing your customers is important to you, use this type of selling venue as an adjunct to other types.

# Choosing Your Selling Venues

Now that we've examined everything from farmers markets to interstate shipping, I hope you're not completely confused. Here's a closer look at deciding how to fit a particular selling venue to your hobby farm.

## Think Locally

The best way to start this decision-making process is to learn as much as you can about all selling venues that you think might make sense for your farm. Certainly, you can start researching the ideas in this book and online, but eventually you must think locally. How will a particular venue affect this little business of yours? What will you bring to the local food and

## Evaluating Selling Venues

Even though customers love the Veggie Van, Denise and her family are still in business to make money. Keeping a close eye on expenses and profits is critical when evaluating a selling venue. So far, the van is working out well.

"As far as going from a farm stand to a mobile farm, the most difficult task we have to overcome is the cost of driving," says Denise. "Right now we visit 21 facilities, but we have had more inquiries and will be checking up on them for the next season." Fuel prices, maintenance to the vehicle, and inconveniences such as flat tires all take their toll on the Veggie Van. But Denise says that she's making enough money to continue.

Still, Denise will spend the winter months considering changes and additions. "Since this season coming will only be my second year, I won't be able to weed out those stops that are less profitable just yet. There's always room to expand, too, but we haven't crossed that bridge yet," she says.

Another aspect to evaluating selling venues is looking at how much competition you have. So far, Denise isn't concerned about that. However, the Commonwealth of Pennsylvania has asked her to help other communities start a mobile farmers market service. "The idea is definitely growing in popularity!" she says.

agriculture scene that is missing or could be improved? Do you understand enough about where you live to answer these questions? If you can't answer them, spend time researching and learning and getting a feel for what it means to be a local business in your area.

Thinking locally sounds simple and seems obvious, but in the 21st century we are inundated with the concept of thinking, acting, marketing, selling, and investing globally. If you've been an executive for a big firm or worked with a company that is foreign-owned, you probably haven't thought locally for a long, long time. In fact, in many work cultures today, if you stood up at the sales meeting and announced that you were going to think locally, your manager would probably blanch and offer you a pink slip.

As rural folks, we're actually challenged with the opposite; that is, trying to understand global markets for commodity products produced on our local soils. It may be difficult for some of us to realize what it means to raise animals or products on a farm and market them nearby. Knowing this, you may find that thinking locally is a place where rural and urban cultures sometimes clash, or at least require a greater understanding of one another.

Sabrina hustles as a large crowd lines up at the Veggie Van's display table.

Direct-to-consumer selling changes all those built-in MBA theories to some extent, and in some ways makes your world seem quite a bit smaller. Don't fear this concept, though, because "local" is a market niche, and it begs to be served. I don't suggest having an isolationist attitude where you stop reading the paper and caring about global politics, but remember that when it comes to serving a local market, you must think about what is best for that niche.

## Questions to Ask About Each Selling Venue

As these thoughts sink in, consider the questions I recommend you find the answers to when deciding on your selling venues. These are the questions you need to ask yourself and the organizers or coordinators of the locations you're considering. Take notes and refer to them. One of the easiest things to do is break down your notes into list of pros and cons.

- Is the selling venue convenient for you and for potential customers?
- Is it attractively maintained and well kept?
- If the location is established, what has been the track record of success? How long are vendors staying there?

- What are the traffic patterns for the area? Is there a peak traffic time or sales time? Will you be there during that time?
- What is the main customer demographic? Is that demographic one you want to serve, and is it one that wants your products?
- What are your insurance and liability requirements? What insurance and protections do the venue offer you?
- What is the seasonality of the location? Does that fit your plans?
- What costs and fees are involved, and when must they be paid?
- Do you have to join a membership organization? If so, what benefits come with membership?
- If you're selling at the venue of a special-interest group or a church, what is the group's mission? Is it consistent with your values?
- What changes does the selling venue require of you? What are the costs of those changes in terms of time, labor, and money?
- Would selling there offer promotional and marketing synergies and advantages for your business?
- How long do you think the venue will continue to exist? What is the longevity of your opportunity to sell there?
- What is the competition from other vendors like?
- What amount of work is required to set up shop? Do you have the time to commit to that setup work?
- What are the labor needs to sell there?
- Would you enjoy selling in this environment?
- Is the location a desirable one for any of the reasons just listed?

## Combining Selling Venues to Maximize Customer Contact

If you've already been thinking about more than one selling venue, great! You're thinking right along with me and the tempo of this book. I can't think of any farm-fresh business that doesn't combine selling venues in its sales plan. It just makes sense when you begin to think of the synergies that can come from using more than one method of selling.

For my own business, I combine four strategies: several types of farmers markets, delivery, customer phone orders, and selling to restaurants. When you're considering selling strategies, also consider wholesale and whether it will work for you as a selling venue.

How you decide which venues fit together comes somewhat intuitively. As you uncover new places to sell, most of the time a schedule seems to fall naturally into place. You'll also be more or less limited in your options depending on the area where you live, your product, and your personal situation. If you don't have another full-time job, you can find more time for additional selling venues than you'd have if your hobby farm is a weekend thing. For people with off-farm jobs, online orders may be the easiest place to start building a customer base.

Here's **a sample week that worked for me during our busy summer farmers market season** from May through October.

**Monday:** Evening deliveries (of weekend orders)

**Tuesday:** Afternoon deliveries (to restaurants)

**Wednesday:** Evening market

**Thursday:** Evening market

**Friday:** Evening market and restaurant deliveries

**Saturday:** Two morning markets (I went to one, my husband to the other)

**Sunday:** Home at the farm

Combining several strategies also saves you time. If you can make deliveries on the way to the farmers market or on the way to restock the consignment store, then using more than one venue to increase your exposure while you're already out and about makes good sense. I recommend mapping out several potential one-week schedules similar to the simple example to the left. Plotting out how different venues will work together over a week's time will help you decide whether they fit or whether your original plan will be too hard to negotiate. You may want to start by committing to things that can be cancelled if they just don't work out. Seasonal markets, online ordering, and consignment selling are all ideas that can be scrapped or changed without a lot of difficulty.

Another consideration is customer demand. When it becomes apparent that many people are interested in buying your product, it may be time to work on your new venture full-time. Customer demand and preferences will also drive your decisions on where and when to sell; you need to be convenient to your best customers.

A word of caution, though: Don't give in to the urge to serve customers' whims too soon. I'm not suggesting you offer poor customer service; I'm talking about making business changes too early in your new venture, based solely on ideas and comments from buyers. It's nice to have customers making requests, but honor those requests, especially any that require you to change your business plan, very cautiously.

## *Juggling Selling Venues*

As I have cautioned throughout this book, you may find yourself running into too many options too fast. When Denise sent letters to administrators at area senior care facilities, she received 21 acceptances! Suddenly, she had a scheduling problem to work out. "I had to make sure we had enough time to drive from place to place and to plan ahead for extra traffic during music fests and other community events," she recalls.

Denise and Sabrina have set times to visit each location a few days a week. Denise's dad, Nathan Gould, also drives a Veggie Van. Right now the operation is on four days a week and averages 215 miles weekly.

Denise works hard to stay on track, but knows that being a little flexible makes her service appealing. "We try to stay at least half an hour at each stop, but we will stay longer if the residents need more time," she says. "I am still making changes and upgrading the operation, so it is difficult for me to say how or where this adventure will take me and our farm. But it is eventful!"

One of the biggest costs and potential disasters for small farm and foodie ventures is growing too big too fast. Believe me, backpedaling, regrouping, and finding cash when you've already maxed out your credit line are very expensive and difficult. Stay the course and simply tell customers that you're happy with their input, but you're just starting out and you appreciate them growing with you.

# *Managing Labor*

Working for yourself has so many wonderful advantages, but plentiful labor is not one of them! As you select and combine selling venues, the amount of labor those options will cost you must be a major consideration. Think about labor in terms of time, individuals to do the selling, and cost—whether that's in salaries for employees or in the value of your own time.

The inevitable labor shortage that comes from you being a one-person show doesn't have to stop you from choosing a selling venue you either really like or think would be really successful. A variety of labor options are available to small farm businesses; you just need to know how to use them effectively.

## The Baardas' Farm Lifestyle

The Baardas love being able to work the farm as a family. Though their various ventures keep them really busy, "I can only say our life has been less than boring or dull. In fact, we are always traveling to other farms along the East Coast to talk and see what they're doing to make their farms work," Denise says. "I am amazed at new ideas and techniques, and we try to implement them to see if they would work for us. Some do and some don't, but at least we can say we tried!"

No matter what the workload, Denise believes her love for what she does helps her maintain a positive outlook. "We realize that all farmers have to deal with time, weather, and the economy, but I think we wouldn't be here doing what we want to do in the busy times and the slow times if we didn't love farming."

Offering shares in your output is a good idea, even if you're not a CSA. With this idea, some customers perform work for you at the farm, freeing up your time, in exchange for a percentage of the crop or food you produce. Another idea is giving customers a percentage or a commission if they sell to their friends and coordinate a convenient delivery or perhaps come out to the farm and pick it up.

Another perfect idea is unpaid internships for high school or college students. You can even offer junior or apprenticeship programs for younger kids. See if your church group would like some team-building or "camp" activities for kids ages 9 to 12, where they get to take home food at the end of their labor. They're not old enough to drive, but if you can keep them (mostly) on task, adolescents with their energy can relieve you of some major manual labor!

## Live-in Labor

No, I'm not talking about your family here—although getting some work out of them wouldn't be a bad idea! What I mean by "live-in labor" is offering a worker a place to stay in your home or perhaps in a guest house or rental house on your property.

This idea has worked well for Aubrey's Natural Meats. In the past, we've offered a college student free room and board and an unpaid internship in exchange for the opportunity to work with a natural meats business. Then I found another farm nearby that needed some livestock and general farm help and was willing to pay by the hour. The student worked probably 40 to 60 hours a week between the two farms, but got about half those hours paid and half compensated by not having to rent a room or buy meals. In this way, the student got both a professional internship and a summer without expenses.

This arrangement works best if you're hiring someone you already know. Consider neighbors or perhaps a nephew, niece, or cousin who'd like the opportunity. At the end of the internship, write up a recommendation letter for the student and ask them to write a short paper about their experience that you can use to attract next year's talent.

## It's All About You!

Here's one of my all-time favorite things about small farm and food ventures: When it's time to go to work in the morning, you choose where you want work to be. Isn't that wonderful when compared to a commute across town? You'd better believe it!

As you consider this chapter about selling venues and the decisions you have to make to select the right ones, remember that it's all about choice. You get to control where you want to sell, when, and to whom. Allow that to be your guide. This new lifestyle of yours is meant to be both fun and fulfilling; don't forget that when you have decisions to make.

### Denise's Best Practices

- Find ways to fill a need in your community.
- Base your pricing on what makes sense for your area.
- Decide what works and what doesn't and stick with the best selling venues.
- Do what you love.

# Chapter 9

# Marketing Ideas

**Learning Objectives**

- ✤ Understand what marketing is and why it's important for local food businesses.
- ✤ Review marketing ideas and marketing best practices for local agriculture-based business.
- ✤ Discover marketing ideas that help promote your product and don't cost anything.
- ✤ Learn the do's and don'ts of calling on chefs from Lewis Shuckman of Shuckman's Fish Co. and Smokery.
- ✤ Get tips for making a sales presentation to retail stores.
- ✤ Learn how to close a sale and get paid in a timely manner.

## Marketing Small Farm and Food Businesses

Marketing is everywhere. There is no product or service, even the generic stuff, that doesn't have a brand or an image or a perception associated with it. Resources on how to market your product or service are almost as prevalent. Marketing books, articles, web sites, and blogs abound, so to say that this chapter is the official last word on marketing is, well, ridiculous. Because so many useful resources exist and because new marketing ideas, concepts, and trends are always being deployed by super-skilled professionals, I also think trying to present the great concepts of marketing in one chapter could be at once incomplete and redundant.

Shuckman's Fish Co. and Smokery has trademarked the name Kentucky Spoonfish Caviar, giving them a unique product and a concept that can be marketed.

With that disclaimer, this chapter will introduce you to ideas specifically for hobby farms and local foods. I'll present concepts that pair as perfectly with marketing local food and farm business as blue cheese with vintage tawny port. A special section at the end of the chapter will help you call on local retailers, and especially chefs. I've added this feature because I learned the hard way that making a sale to a busy chef requires a special blend of patience, persistence, and knowledge of his or her business and needs. Successfully presenting your products to retailers for sale through a store also requires finesse and the ability to showcase your own brand or identity, so I'll cover that, too.

As I introduced in chapter 8, marketing is the other part of the twin towers, along with sales. The amount of time and effort you spend on marketing directly correlates with your desire to get the word out about your business and recruit more customers. Notice that I did not say the amount of marketing you do correlates with your dollar volume in sales. There is certainly a relationship—in some cases a very strong relationship—between marketing and sales, but the specifics of that relationship depend on many factors.

Numerous small farm businesses find themselves almost overwhelmed with customers, even though they do very little marketing. It's true that word of mouth is one of the biggest marketing tools you'll find when you're selling direct to consumers. If you're bringing something novel and nice to an area and you get a loyal group of customers behind you, it's certainly possible that you won't need to spend much time on marketing.

The amount of effort you put into marketing also depends on how busy you want to be and how much income you want to make. If you're reading this book strictly to get ideas for a hobby rather than a small business, you'll likely spend less time developing a marketing plan and more time working on your golf game. Differences in goals and needs are what make marketing tactics unique to each individual company—and that's just fine.

Every new enterprise has to start with some marketing to get the word out, but there are times when you need to market more and times when the act of selling seems to be all the marketing you need to keep busy. Marketing efforts are especially called for when you're bringing out a new product, when you're moving, and when you add anything new to your business—be it partners, animals, or varieties of vegetables. Changing or adding a selling venue is also a very important time to put concentrated effort into marketing.

You'll want to market to your customers when you need to boost sales as well. Sometimes business flourishes, and other times, even if you're sure you're doing everything right, it struggles. This downturn often hits in the second or third year for hobby farms and local foods businesses, because the newness has worn off and some customers have moved on to the next new thing. If you find yourself in this situation, marketing your farm's uniqueness is the way to lure old customers back and find new ones.

# What Is Marketing?

Marketing seems to follow us everywhere we go. If you're looking to move to the country and the sanctuary of the simple life, you may disappointed to hear that marketing, branding, and image will follow you like a thunderhead even out to the new farm. Don't despair, though; marketing for hobby farms doesn't have to mean abject commercialism. On the contrary, marketing for this niche should be as comfortable and natural as the products, animals, and foods you produce.

Marketing for a hobby farm has a softer edge, one that should tickle the fancies of your customers and pull at their heartstrings. Bringing folks to the farm to watch baby calves frolic in spring grass is a form of marketing your new business, but it sure doesn't feel like the marketing you're used to, does it?

For you in a small farm enterprise, marketing means creating customer awareness and a positive perception of your product and then turning that awareness and positive perception into sales.

Marketing is about creating a message that, if repeated often enough, becomes new knowledge. For example, have you ever met anyone who asks for a Coke even if they're actually planning to order a Mountain Dew? The word *Coke* has become the general term in their mind for a sugary carbonated beverage. To that individual, Coke is not just a brand-name product, but an actual category of beverage.

Marketing is the action that turns a brand name into a generic noun. You can do this, too. Even though comparing the new acreage you just bought to a colossal brand seems ridiculous, it's actually not. You can work within your own local

**How is marketing different from advertising?** Marketing encompasses an entire set of activities, advertising being one of them. Advertising is the act of presenting something for sale or presenting a price or a discount to the public. Marketing includes everything you do to represent yourself in the world, including advertising, your web site, your business cards, and even your presence at community events and business functions. Once you start a hobby business, it is almost impossible not to be marketing yourself every time you are in public.

area with your customers to turn the word *farm* into something that always means *your* farm. You can market successfully enough to your customers that when they say, "Let's go buy a steak," they always mean a steak from you. Likewise, when a customer says, "Let's go out to the farm," they mean your farm every time.

# How to Develop a Marketing Plan

A marketing plan is just like any other planning tool you use for your business. It's a set of goals and steps that you create and implement so you can make a concentrated effort to get your message to current and potential customers and translate that message into cash. Marketing plans do not have to follow any textbook design, although every plan should contain the following elements.

- Target
- Message

- Timing
- Cost
- Delivery
- Feedback/analysis

When you make your plan, consider each of these elements. If you're new to marketing or just aren't sure how to market your new food venture, write out a description, specific to your goals and needs, for each element.

Marketing plans don't need to be drafted by any professional other than you. A simple word-processing file will do, and some accounting software programs have built-in templates you can use. You'll find a sample marketing plan in the appendix, contributed by the University of Tennessee Center for Profitable Agriculture. This plan shows one year of planned promotional activities for a retail meat market and a sample budget to accompany it. Once you've made a marketing plan and begun using it, save it to refer to later. When you find a method that's successful one year, you may want to use it again and again.

## Target

The target for your marketing is the group of customers and prospects you want to reach with your message. You probably already defined this group

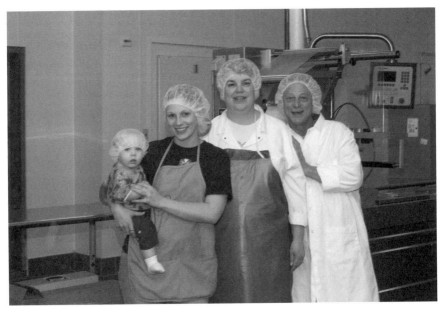

The Shuckman family: (right to left) Lewis, Vicky, Lauren, and Brennan.

broadly when you did your business plan, your market analysis, and even your choice of selling venue. Now you're going to come up with the message you want these customers to hear.

## Message

The message is the information you want to put out there. The message can take many forms, and is likely a combination of image and text. The important thing about message is to use the chosen medium (print, Internet, radio, and so on) to get the result you want. For example, if you want to reach a wide and varied audience that will likely find you via an online search, print advertising in the newspaper is the wrong route; you need to seek Internet ad venues. Or, if you have a very local market and want to promote a special during an upcoming festival, the local paper's Sunday metro section is probably a good choice. Even with those examples, however, most marketing plans call for a combination of media because people get their information from a variety of sources.

## Timing

Savvy marketers and professional executives know that the timing of the message is very, very important. The customer has to receive the message at the right time to be convinced to buy. Examples can be glaringly simple: Don't bring out all your marketing efforts in the winter if you operate only during the growing season. Start marketing more heavily when you're ready for pre-orders or when the weather starts to warm up and people begin to think about coming to the farm or to the market again.

The timing of your message also needs to be convenient to you. If your business is seasonal, for example, you don't want your message to arrive too early or you risk not having enough product to fill orders. Likewise, if you wait too late in the season to advertise your fall open house, your customers may have made other plans for that date.

So to sum up, timing must benefit the customer and work with your cash flow needs and product availability.

## Cost

Cost is important, of course, because how much you spend on marketing activities affects your bottom line. A strong marketing plan includes a cost-benefit ratio. To calculate this ratio, decide how much money in sales you want in return for the effort and price tag involved in marketing. Some part of this number is completely unknown even to seasoned marketers; the best

## Meet Lewis Shuckman, Producer
## of Kentucky Smoked Fish

"This is Lewis Shuckman, how you is?" is how the president and pro-prietor of Shuckman's Fish Co. and Smokery answers the phone. The cheerful, if not grammatically correct, greeting provides instant warmth. Lewis puts people at ease.

The native Kentuckian has the food business in his blood. Lewis's grandfather started a meat market and grocery in 1919. By 1954, Lewis's father took over and moved into wholesale meats and grocery. By the late 1990s, with Lewis involved, the company began to add fish and seafood. The business has remained in the family, too—Lewis's wife, Vicky, and daughter, Lauren Belbenger, work in the operation.

But some things have changed. "We started to take a real hard look at acquiring and smoking fish," Lewis recalls. "Back in 2000, we got out of all other proteins except fish."

He says that choice was due in part to economics, but also had to do with staying current in the marketplace. "As a family business, you have to make rational decisions. We knew American eating hab-its were changing, and we saw a chance to get into aquaculture in its infant stage," Lewis says.

Aquaculture, or the practice of farm-raising fish, has been grow-ing in popularity among farmers as a way to diversify. Lewis liked the

you can do is forecast expected return on marketing investment based on the knowledge you have and the research you've done.

For example, say you want to earn $10,000 your first summer in busi-ness. You have to reach enough people to translate those contacts into the $10,000 in sales.

Let's say it will cost $100 per season for a spot at a farmers market, so that is a marketing cost. Having flyers printed and distributed to a variety of target locations will cost another $100, and having a quality web site set up will cost $3,000. So that's $3,200 in marketing costs.

Now you need to decide how many people you'll reach with the mar-keting efforts you have planned and how many people it will probably take to reach your sales goal. Exceptional salespeople expect that only 30 to 40 percent of the contacts they make will result in sales, and I think that num-ber is a bit high. It's more likely that about 10 to 20 percent of the people you contact will become actual buyers.

idea as a means to support Kentucky farmers, grow a Kentucky product, and market something unique.

Aquaculture seems to work well in Kentucky; the clear streams produce wonderful fish in a clean environment. "In Kentucky we have more free-run springs than any other place on earth, and we have some of the finest lakes in the world; they're clear, pristine, and very natural," he says.

"Aquaculture was very young, and we saw an opportunity for producers in Kentucky," Lewis continues. He initially encouraged small farmers to add fish production to their businesses, and he continues to work with and visit farmers today. "Aquaculture requires little investment compared to other agricultural businesses," he notes. Of the people he talked to about fish farming, "Some scratched their heads and said, 'Well, I don't think so,' but others said, 'Well, you may have something there.'"

Right from the start, Lewis was thinking about marketing. "I knew we needed to get consumers interested. So we first relied on freshwater species, trout and spoonfish. As times moved forward we added salmon," he says. "We grew the product line slowly. Our goal here was to help Kentucky producers and offer chefs and stores a unique product."

So if, for your $3,200 spent, you reach 1,000 people in the local newspaper and with hits on your web site in the weeks leading up to the farmers market, then at a 20 percent conversion rate, you might bring in as many as 200 new customers. That means you spent $16 to get each new customer. If your average item costs $1.50, then 200 new customers may not be enough. However, if your average item costs $10, you're getting closer.

Still, as far as setting up a web site goes, that cost can be spread out over several years, so you don't have to be as concerned about recouping it immediately as you would have if you had spent $3,200 in newspaper advertising for one event or festival. For a sample advertising and marketing budget, see pages 262 and 263 in the Sample Business Documents.

Your prices also figure into your cost-benefit ratio, which is simply a comparison of the costs for the amount of sales you generate. For higher-priced items, you'll need fewer transactions to reach your goal. That means your marketing may need to be more highly targeted to get to the people you think will most likely pay the price you're asking.

# Uniqueness as a Marketing Tool

One aspect of marketing is to explain to customers why your products are unique and special. Shuckman's Fish Co. and Smokery has done that by offering a line of Kentucky-only products. To meet demand, Lewis must acquire fish from all over the world, including Canada, Chile, and the East and West Coasts of the United States. But his made-in-Kentucky products account for more than 10 percent of his raw fish use.

The signature product in this line is Kentucky Spoonfish Caviar, a name Lewis has trademarked. "The American Paddlefish, also called the Kentucky Spoonfish, is Kentucky's own identity," Lewis explains. "This gives Kentucky another icon, along with the Thoroughbred and bourbon industries."

Kentucky Spoonfish Caviar has become an important part of Lewis's product line. "Chefs love it; they'll tell you it is to die for! It's a beautiful thing in that it is an American product and it's a freshwater product," Lewis says. The caviar comes in two- and four-ounce tins.

The Kentucky line includes four varieties of smoked fish spreads. Five-ounce tubs are sold to consumers; chefs can buy in the five-pound size. Other products include Nova Smoked Salmon, a cold-smoked salmon that Lewis says is similar to prosciutto, and Alder Salmon, a hot-smoked item.

Black Mountain Smoked Trout is another unique Kentucky product. "The reason it is called Black Mountain is because I was working with these folks from eastern Kentucky and there are several closed mines in the area," Lewis explains. "Clean, pristine water that is cold all year comes down from these mountains. They make a gravity flow feed to the tanks and grow the fish in it." Lewis likes the fact that he is able to bring business to people in the old coal-mining regions of his home state.

Lewis manages to add a local element even to the fish he brings in from outside. "We cure the fish in ten-year-old handmade bourbon barrels," he says. The barrels are dismantled and the hickory wood is added to the smokehouse. "Outside, the aromatics make you want to eat it!" he says.

## Delivery

Delivering your marketing message means getting it out there in front of the public. The most effective way to reach your target several times is to use a variety of delivery methods.

You can deliver your message in person, such as when you're selling at a farmers market. Or the setting can be more formal, such as a talk about your business at a Rotary International chapter meeting or a Lion's Club. Delivery is also accomplished through media such as newspapers, radio, magazines, and television. You have no control over these media, but you do have control over delivery on web sites, brochures, flyers, and business cards that get your name out there.

## Feedback/Analysis

After you've created your marketing message and it has been out there for a few months, it's time to assess. Never skip this step, because doing so can result in your

Lewis's processing facility is modern, but personal attention to detail and quality control still prevails.

new business being something akin to the Israelites' forty years spent wandering in the wilderness: You'll be going around and around the same mountain, wondering why there is no new result.

Gather your feedback from customers and peers, as well as from your own perceptions. Give customers a discount or a coupon if they will tell you how they heard about you in the last few months. Ask peers you trust and respect if they saw your marketing efforts in the local area, and if so, what their impressions were.

You don't have to take every comment to heart. A competitor with an axe to grind isn't going to be any help, but a collective evaluation of responses and your own evaluation as it relates to sales is essential. This is a chance to examine your cost-benefit ratio, or the relationship between the cost of marketing and the actual benefit in terms of money you received from the effort.

## How Often Should Marketing Plans Change?

Marketing plans should be made either in anticipation of events and activities or to achieve a desired result. So it almost goes without saying that you need to review your plan as often as it makes sense for your business.

## Feedback from Customers

Listening to his customers is one of Lewis's most effective marketing strategies. "You'd be amazed at what you can learn when you listen to what people are saying—you can really increase sales," he says.

One customer suggestion yielded an increase in business. Lewis was talking to a single woman who thought his fish packages were too large. He downsized from a four-serving to a two-serving package and ended up selling more packages of smoked fish overall.

Gathering outside input is an important factor in Lewis's marketing strategy, and he frequently talks to peers and customers—at both the retail and the restaurant level—to make sure he is delivering what they like most. "I suggest to people that they hire a business consultant; they can save you a lot of money," he recommends. Consultants give him an objective look at his business, enabling him to see areas of inefficiency and discover new ideas.

For example, if your business is seasonal, take a look every year after you've closed up shop or stopped production for the season and evaluate what marketing efforts worked and what needs to be improved or added before you start up again. If you run a year-round venture, review your plan at least every six months and decide whether you're on target or you've gotten off track and need to make changes.

Either way, take a critical look at your plan at least once a year. Examine your accomplishments and the trackable sales benefits that occurred as a direct result of your planned marketing efforts. Evaluate any indirect benefits that occurred as a result of your marketing efforts, too, such as an opportunity to network with other farm marketers or newbie farmers and what you learned from those experiences, and assess how those things will result in tangible benefits (sales) in the future. Finally, decide where you fell short or what parts of the plan you didn't stick to and consider why you let those slide.

## Marketing Ideas

Marketing is a series of activities. Your job is to glitz up the old flyer–brochure–business card–web site routine. That said, if you haven't done it already, those four items are probably a necessity in today's market. You don't need to spend a lot of money on these basics, although you certainly can. You can also find a lot of free graphics and ideas for these minimum marketing items by browsing the Internet.

My main advice—and this comes from experience—is that you start with smaller quantities rather than opting for the discounts that come with ordering large amounts. The reason is simple: You'll make more changes after the first year in business than you will in the next several years combined. I have stacks of old brochures—professionally printed, of course—with old e-mail addresses, logos, and selling locations that I realized just didn't work out after my first year in business. Save yourself some money and go light on the quantities until you've found your rhythm.

Here are some marketing ideas that are basic and some that are more creative. But remember, the only effective marketing plan is the one that is specific to your business.

## Logos

One of the best ways to market your business is through a physical image such as a logo. I love logos because they are so personal. Create your own or obtain professional help in creating a visual image that expresses your concepts and feelings for your business.

What should be on the logo may be obvious, or it may take you months to come up with the right image. Let it take as long as it needs to; remember, this isn't a high-pressure corporate environment, and your marketing should reflect what you want your small farm to be.

Logos, once designed, can be put on just about every object you can think of. Find excuses to get that logo out there every day.

## Local Sponsorships

Marketing is one of the ways you endear yourself to the consuming public, so go to the causes that matter to get yourself seen. There are more opportunities to sponsor organizations than there is money to do so. Your contribution will not only be appreciated, but it's needed by someone.

When you sponsor something, the more local the better for the highest impact. For example, in rural areas, sponsoring the local Future Farmers of America (FFA) chapter is always a hit and offers a much-needed support system. Consider helping to sponsor the local group's trip to an educational venue they've been wanting to visit or just helping out with their annual fundraising drive.

Another idea, especially nice for more citified areas, is to help sponsor the development of an urban or community garden. Perhaps you can

Logos are a very personal way to establish a brand identity on everything you sell. This is the logo for my meats business.

contribute seedlings, time, or funds to help install the season's crop. Find ways that have relevant impact for your business but are also passions for your heart.

## Gifts and Favors

Consider giving out freebies that reflect your image and personal style to customers who purchase your product or even just sign up for your e-mail list. These gifts serve as a reminder of your business and are a token of your appreciation for your customers. Of course, all gifts should have your company name and logo prominently displayed.

To avoid ending up on the floor of a busy mom's SUV, make sure the item you give has regular use and is truly appreciated by the recipient. If you sell at farmers markets, consider purchasing and giving out reusable shopping bags or cooler packs with your logo on them. For winter markets, why not offer mittens or packets of hand-warmers? If people come to the farm, give them a dried flower bouquet or something characteristic of your place.

# Marketing Ideas That Are Free!

Advertising is one of the functions of marketing, but let's face it, advertising costs plenty. I'm not suggesting you don't advertise if you feel it is your best marketing tool, but for many, the cost of advertising and the fact that it may not reach your target market make it a poor choice. For those of you on a tight budget, I present a handful of my favorite free marketing ideas. All of these have worked for me at one time or another.

## Volunteer

One of the best ways to market yourself and better yourself at the same time is to volunteer for a cause that matters to you. Being out there representing yourself as an owner of a small local foods business and helping others creates a positive image of you that will last longer than any ad campaign.

Other ways to volunteer include becoming a member of local civic, political, or social organizations and serving as an officer in such a group. Volunteerism is completely personal, and so many kinds of groups need volunteer help that ideas are limitless. My husband and I have helped coach livestock judging teams and even hosted school FFA football games in our pasture. There's a unique volunteer opportunity out there for everyone.

Volunteerism should be combined with several other marketing tactics, because the narrow focus and the amount of work typically required means it won't ever make you rich. But it will lead to some sales and make your heart happy.

## Piggybacking

The idea of piggybacking is that it's a free ride. That grade school game is just as appropriate to market your small farm or local foods business. Find someone who is doing marketing anyway and ask if you can add your material to theirs. For example, ask your farmers market to distribute your flyers in their mailings (they usually will if all vendors are allowed), or post cards on other people's bulletin boards (where it's allowed or invited).

Sometimes marketing tactics are as simple as doing what is right. **"If you put your name on it, it should be the best you can produce,"** Lewis advises. "People appreciate consistency, and that's how you get repeat customers."

## E-mail Newsletters

Keeping in touch with customers, prospects, and other interested parties is a great way to get free marketing. Newsletters sent by e-mail work because you are providing free information to those who have asked to receive it. With regular contact comes more regular sales from your core customers. I cover newsletter content in more detail in chapter 10, but suffice it to say that you can easily create an e-mail customer database on a home computer and then regularly send your own correspondence using a simple e-mail message or even a free newsletter template found online or in your word processor.

## Blogging

If you like to write or feel you have something to say, blogging may be the best idea yet for free marketing of your new venture. Most blogs can set up free of charge simply by signing up with a provider, such as ProBlogger or WordPress. The simple keys to blogging are regular updates and adding value with your content. For example, if you are a meat producer like me, blog about how to cook a certain cut of meat or how to prepare a recipe. A beekeeper might blog about how to keep honey from crystallizing or how to make a delicious marinade.

Of course, you must also let people know that your blog is out there. I suggest getting several quality posts ready to put online before you introduce people to your blog. That way, they'll get a representative sample and some good reasons to come back.

## Media Coverage

One of my best secrets to free marketing is using the media to market for you. Basically, you call up your local newspaper, foodie or gourmet magazine, green living magazine, radio or TV station, or any roving reporter and ask them to do a story on you. Sound arrogant? Honestly, it's not. If you've got a great story to tell, the media will be interested. I've been interviewed in numerous media outlets, from the *New York Times* to the local *Elwood Call Leader.* (The difference in demographics and circulation here is so dramatic, it's comical!)

The key to using media effectively is in the angle your story takes. In some areas, the simple fact that you've started a local foods business may be enough—it was in my home area of Indianapolis—but in areas where the market for local foods is more saturated, you need a unique take on the theme to snare the media's attention and garner your free press. Think about how you're different, special, or quirky.

Have you saved or resurrected a small farm from destruction, updated a historic building into a shop, or formed a cooperative? Have you started something completely new to the area, such as a local foods delivery service or a CSA? If you're the first of your kind in a market, then you're a news story. If that is not the case, pitch a story about how you contribute to the growing local foods movement and consider partnering with other vendors to pitch a whole concept to the media that will feature a group of like-minded local foods entrepreneurs. Always think of your special angle—it is your hook for the story.

A related idea is to see if you can place a free informational article or be interviewed by the newsletter or journal of any trade group you have joined. You may find that marketing to those with similar interests has benefits that make it worth the time. Marketing to those with similar interests helps because it can create networking and partnership opportunities. It also helps when you can post that article on your web site or hang it up in your store or even e-mail a copy to your customer list to establish you as an expert with both customers and peers.

Using the media is not without a host of caveats, the most obvious of which is being misquoted. You can ask to preview the work before it goes into print, but the journalist or editor is under no obligation to do so. Another problem can arise if you're too honest, too casual, or not particularly well spoken. I don't mean to insult you, but, hey, it's true of a lot of people. If your remarks make you sound less than desirable, that image is what your potential customers will see. Be careful and professional when you are being interviewed!

I've found the final little drawback to being interviewed is the hardest to prevent. Let's call it the "Be careful what you wish for" clause. I was once interviewed by a farm-oriented newspaper in my state. First of all, this wasn't a very good venue to be interviewed in from a marketing standpoint because the readers were all people like me and were not planning to buy from me now or ever. Second, the writer slightly misquoted me and put that in the large heading! The heading read something like "Farmer Seeks Fellow Cattlemen to Supply Beef for Natural Meats Biz." This was just wrong. I mentioned to the interviewer that at times my demand exceeded my supply and so sometimes I had to buy beef from neighbors to fill orders. I went on to say that those neighbors had all been prescreened and I wasn't seeking any additional beef producers to buy from, but none of that made it into the article.

I ended up with my phone ringing off the hook with solicitations of all kinds. I got 32 calls from all over the country (people had apparently read the article online) asking me to buy their beef. I had practically as many calls from other would-be natural beef producers wanting free advice on how to start a meat business. There were also people who raised lambs, goats, chickens, rabbits, and pigs, all calling to see if I'd care to buy their animals and branch into another meat. Plus, I had sales reps peddling everything from feed to fertilizer, mineral blocks to web design. I actually received one call from a poor guy in North Dakota who wondered if I'd like to buy his 27-year-old metal bumper hitch trailer! Oh yes, be careful what you wish for!

Frankly, preventing some solicitation can be difficult when you put yourself out in public. But to avoid the ridiculous barrage of calls that plagued me for weeks, select media outlets that you think are truly going to be read by potentially interested customers. My favorites include gourmet sections in the newspaper and foodie magazines.

## Prospecting Local Retailers

Prospecting for new business means doing work to set up a wholesale or retail account, or setting up a consignment deal with a shop you're interested in getting your products into. This is one of my all-time favorite small business activities. You may have just read that and slammed this book shut. If you're still reading, you'll learn that prospecting doesn't have to feel like you've become the annoying salesperson at the discount mattress store. Prospecting in small farm and local foods business is one aspect of marketing; it should fit as naturally with your style as everything else you do.

You want to make a nice presentation and get the prospect to want you, but you are also interviewing them. **While you are there, you need to find out whether you really want this opportunity.** Be sure to have a set of questions ready so you can acquire and later dissect enough pertinent information about the business for you to make an educated decision about whether this is the right fit for you.

## Preparing Your Presentation

Selling Skills 101 comes into play here, but let's review a few of the basic items you need to prepare as they relate to foods and farms.

- Call first and ask for an appointment, and then call again the day before to confirm it.

- Prepare and rehearse a short description of your business and yourself. Be as unique as possible and tell your story in an engaging way.

- Make sure you can present your product and explain it clearly in less than two minutes.

- Make your message unique for each location, based on what you know about them, while maintaining honesty and the image you want to present throughout.

- Have items to leave behind, such as a few brochures, business cards, and price lists.

## Making the Prospecting Call

The first order of business is to arrive on time (unless there is an emergency). Consider rescheduling if it appears the person you know is the buyer hasn't shown up or will be exceedingly late. Leave plenty of time during the meeting for questions, conversation, and ordering. Don't talk the whole time and leave them with a weak "Well, just give me a call if you're interested." This approach is 95 percent likely to fail.

You may or may not wish to leave behind samples for those in charge to try out. Don't feel like you have to be too generous with samples. I've seen prospects who are used to big companies calling on them, and they can cash in on the freebies. Don't feel obligated to leave a sample unless it will solidify the deal and you really want to work with this retailer.

## Calling on Retailers

Lewis knew he had a unique product that would set him apart in a retail display case. Getting that product into retail stores was the first task, and Lewis started with cold calls. "I'd call and make an appointment," he says. When he got there, he found several things were important in making his presentation.

First, find out how much space the retailer has in his cases. "Don't go in and act like, 'I need five feet or eight feet or whatever.' Find out what is available," Lewis suggests. He also recommends that you have a frank discussion with the retailer about how much money they need to make. Although pricing is personal and you also need to make money, Lewis says blending into the general price range at the store is important when you're starting out. "Fall into the same price structure or be within the same area of similar products," he advises.

"One thing I found stores liked was that they could call and place an order and I could get it to them that afternoon," he recalls. Eventually, good business led to referrals, but it was two years of cold calls that got him started.

Finally, tasting and events drive the food business, so make it clear to the retailer that you intend to do what it takes to encourage business. "Nine out of ten people won't grab something without trying it," he says. So he decided to offer in-store demos—a marketing strategy he continues today.

"Be there to support your products. Provided you do that, the customers will come back," Lewis says.

## Meeting the Customers

Once you've gotten your product into a retail location, you are not done marketing. In fact, there is more work ahead to ensure that the selling venue continues to pay dividends. Most local retailers will let you host special events at the store (with them involved in the planning, of course) and may even share incidental expenses with you.

If demos are not an option, plan to be on hand to meet customers on busy days or on days the retailer is already promoting as special events (this is piggybacking, of course!). For most buyers of local agriculture products, it is the connection with the face and hands that prepared their food that will guarantee a purchase.

# Prospecting Chefs

Before I close this chapter, I've just got to tell you that calling on chefs is fun, interesting, and rewarding, but typically not easy. When I first started Aubrey's Natural Meats, I thought I wanted to sell to chefs exclusively. Though it wasn't easy, and I didn't know it at the time anyway, I cut my small farm business teeth on the most discriminating customers—chefs.

This section offers my best tips, all gleaned from the pretty and painful experiences I had in serving the food service industry.

## Preparing Your Presentation

Prepare for your presentation to chefs in the same way that you prepared for retail prospecting calls. The major addition for calling on chefs is that you must bring samples. Make sure the samples are attractively packaged, are clearly labeled with the product name and/or description, and include your name and logo. Bring as many samples as you think you can fit into a short presentation. Make sure these are the items you're looking to sell, rather than your whole product line.

## Making the Prospecting Call

I found one big difference between calling on chefs and calling on retail stores: More than any other type of client, chefs want to know, "What's in it for me?" Part of this is a function of their jobs—they are credited with the great meals and chewed out for the not-so-great ones. There is no passing the buck when you're talking about a person's satisfaction, or lack of it, with their dining experience. So you must make sure the conversation you have with the chef is full of references to how your product is unique and special. You need to make that chef feel as if he or she just has to have the item to offer diners the greatest possible food experience.

Think about your production and inventory, and don't bring samples of your top-selling product to every chef unless you make a *lot* of it. **If you can't fill the orders, you'll start off a new relationship with a restaurant by letting them down.**

Another tip for calling on chefs is to remember that they are all foodies. I almost always asked the chef to prepare one or more items from my sample kit while I waited (remember, I sell raw meat). If the chef liked it (many chefs have large, easily offended egos regarding their palates; walk softly around their opinions on taste, smell, and texture), I'd invite other staff, including sous chefs, servers, and other managers to try the item, too. Believe me, oohs and aahs from staff sway the chef—whether he or she will admit that out loud or not.

## Timing Chef Prospecting Calls

When you call chefs for your initial appointment, timing is essential. You simply cannot call a new prospect during prep time or when meals are being served. Some chefs will do you a favor and not answer their phones; others will curse you and your future generations if you interrupt them during a busy dinner service. Best practice: Do not call within an hour of the lunch service (typically after 11 a.m.) and within an hour and a half of the dinner hour (so not after 4:30 p.m.).

When you get to the restaurant, don't be surprised if you do not get the chef's full attention during your appointment. It is not uncommon for you to be making your presentation while the chef is chopping vegetables, disciplining a sous chef or server, and answering you with barely polite "uh-huhs" while not looking up. This doesn't mean you're not getting through. It's just that a chef's daily tasks are always time-sensitive; when patrons come into the restaurant, the food must be ready!

Ask for confirmation that the chef understands and is interested. And again, offer to let the chef cook those samples. Tasting is believing!

Lewis enjoys converting customers to the fine but acquired taste of caviar.

## Setting Up Payment

With retail stores, payment collection is usually straightforward; your biggest challenge is negotiating how soon you get paid. Knowing whom to bill and collect from in a restaurant can be more complicated. You need to gather some basic facts about how the business is run. Is billing handled by

## Creating a Market for Caviar

Lewis says that while American diets include more and more fish, many U.S. palates are not used to caviar. He enjoys converting consumers, and starts them out easy. "At the entry level, for folks who are not big fans of seafood, we start them out with the spreads, then move to the smoked fillets, and then we try to move them up to the caviars," Lewis says. "We like to introduce new things to people."

He believes people eventually come to enjoy his caviar because it is approachable. "Our caviar is available, affordable, and it is not snobby!" He enjoys seeing more and more young people enjoying caviar without being concerned about the status of imported varieties.

To begin selling his smoked fish and caviar, Lewis started with fine-dining establishments. The sales went well, but real growth came when he began to get to know the servers selling to customers at the table. "I learned that I need to get in there and spend time with servers. I needed to let them taste it and explain it to them," he says. "If the servers don't support the chef, you can't accomplish your goals." He knows that patrons listen to their servers and that suggestive selling in a restaurant really works.

an absentee owner who is not a chef, a chef-owner, a chef who is not an owner, a restaurant manager, a general manager for that division of a larger business (such as a country club), or an accountant who is not involved in operating the restaurant at all?

If you call on restaurants, even in a small town, you'll come across every one of these bill payers and more. So you must always remember to ask who makes decisions, who needs to be copied on orders, and who pays the bills. Sometimes this will be three different people!

My experience is that if you don't ask these questions and you give a bill to a busy chef who simply waves you away with, "I'll take care of it," you are setting yourself up for a long wait for payment—or no payment, because your bill has been lost and the evidence, of course, has been eaten!

## Customer Service for Chefs

As with retailers, offer chefs the opportunity for you to visit and meet with customers. Some chefs will love this idea and will be delighted to have you

# Calling on Chefs

Lewis says that calling on chefs requires its own brand of patience and understanding. Though he now uses a distributor to sell his products and handle the delivery, he still maintains a working relationship with many chefs and rides with the distributor on sales calls. He offers several tips for working with chefs and gaining their business. "First of all, chefs have a price structure on their menu and they need foods at a certain price, regardless of whether the restaurant is upscale or casual," Lewis says. "You have to listen to what they have to say about what they can spend. Have a product that fits into that window."

Another key in working with chefs is to help them get the per-portion cost down and still deliver a high-quality dish. This also applies if you want to work with caterers.

Lewis recommends being fairly forthcoming with chefs and letting them know details about your operation. Many will want to visit. "Just about all chefs love to do the local thing. They come down here all the time to see what we're doing," Lewis says.

And finally, be persistent yet considerate. "Don't assume that a chef is going to carry your product," he says. "And if a chef is just slammed for time, set another appointment and come back later. You want to make sure they can go through things with you."

Some chefs need time to get to know you before they are ready to add your products to their menu, Lewis says. It can take months to convert one chef over to your goods, while others will buy on the first visit. "If he don't know you from a lump of coal, a great way to start is to try a weekend special and see how it goes."

in their restaurant promoting your locally raised products. But not all chefs will want to share the glory. Many chefs prefer to be the expert, and if you come in and explain everything about the process and the foods, you may step on their toes. So offer this idea humbly or not at all if you sense strong territorial behavior from your chef.

I would insist on tabletop information about your products, particularly if you also want direct to consumers to your on-farm business. I had a menu insert brochure prepared and printed. To make the chef and restaurant feel exclusive and special, I had their name printed on the top, such as "Handley's Steakhouse Presents Aubrey's Natural Meats."

Finally, set up regular ordering and delivery. Again, make sure you know both who takes the orders and who pays. After that, you need to talk

## Introducing New Products

Being in the food business, Lewis knows that he is subject to the fickleness of consumer trends. He continues to grow, improve, and change his product line to remain unique. When considering a new product, he says, "You need to ask customers if they think you're crazy or not. That's just the way it is!" He also researches his marketplace so he can be aware of new food products and what's coming to his area.

Though change has been good for Lewis, he isn't in a hurry to make changes, preferring to give products time to prove themselves. "Don't be out of the box and into another box, just to be outside the box," Lewis advises. "Just because you are the entrepreneur doesn't mean you're the smartest person on earth. You still need to listen and learn."

Lewis is also an avid experimenter with his products and tinkers with blends of smoke and types of hickory wood. "Never quit making mistakes," he says. "Just don't step in the same hole twice."

Sage advice!

with the buyer or chef about how frequently they prefer to place orders and what time of day they want you to call them to discuss it. Also, establish your delivery practices, including time of day and day of week they prefer delivery and where you should bring your product.

# Stay on Task Even if You're Flooded with Tasks

Marketing is a broad topic, and making prospecting calls for your own local foods business takes persistence and a thicker skin than you might be used to. Still, this is your new venture, and you should feel proud and confident. The knowledge that you grew this or raised this should shine through in every marketing activity. Oh, and by the way, that pride in itself is a kind of marketing!

Don't forget marketing when you get busy. I know I mentioned earlier that sometimes you truly need marketing, and I still hold to that, but you also need to keep seeking the next sale and the next customer no matter how good you've got it today.

You won't find Lewis sitting around much, but here he takes a break from his schedule to catch up on paperwork in the office.

During busy seasons and times of prosperity, marketing can be the last thing on your mind. So here's my advice: Assign the task of marketing to a responsible person. If that person is you, block out time for marketing tasks on a calendar, and make sure you're doing it weekly or biweekly. Most electronic calendar programs offer a reminder feature, so set that to pop up once a week so you can remind yourself to remind yourself.

If you're not good at marketing or suspect you'll be the busy person who will likely forget, perhaps assigning the job of marketing is a way to involve a spouse or even your parents. If you give those interested peripheral individuals a task that doesn't require them to leave the house, they can stay interested and really help you out in the process.

## Lewis's Best Practices

- Let employees do their jobs; don't micromanage.
- Listen to vendors of your raw materials and buy quality.
- Update your financial and insurance documents every year.
- Consider hiring a business consultant.
- Let customers and employees know you appreciate them.

# Chapter 10

# Keeping (and Keeping Track of) Customers

**Learning Objectives**

- Learn the importance of building a customer list.
- Get tips on turning that list into a useful database that is easy to use.
- Get ideas that will help you retain customers.
- Learn how to create and distribute a customer newsletter.
- Learn how to handle customer returns and complaints diplomatically without giving away the farm.
- Discover how to get current customers to bring in new customers for you.
- Meet the owners of Raven's Glenn Winery and Restaurant.

## Customers Are Your Most Important Business Asset

We tend of think of assets as tangible—things we can feel, touch, and own. When you're preparing a balance sheet for the bank to examine your credit, you'll likely list everything from property to household possessions. Your first day in business, those are basically your only business assets. Yet before long (assuming you're seriously following the steps found in this book!), you'll own another asset—a customer base.

If you've never worked for yourself before, you may not realize that customers are a part of your business's assets; you'll figure it out quickly when transactions and cash start to flow in your direction. As far as the balance sheet goes, a customer base can eventually be listed as an asset, even though, unlike vehicles, there is no official Kelley Blue Book–assessed amount.

Now that you know that customers and the sales they generate are positive line-items on your business documents, consider just how essential an asset your customer base is. I'm willing to say that your customer base is your single most valuable asset, even though you can't see it every day or keep it in the barn where you "know it's there." That's why you've got to nurture and protect your customer base just as you would the other important assets on your hobby farm. Taking care of livestock seems natural because the little mouths beg to be fed every day. Equipment maintenance and updates are also necessary to keep the gears running smoothly. Hard assets require upkeep, and you'll notice when you haven't kept up your end of the work. (Believe me, it would be impossible to ignore the bellowing cows around here every evening at 5 p.m.!) But the care and feeding of "soft" assets—the intangible ones—might not be so obvious.

**To make sure all the tangible and intangible assets are included in the value of a business,** many professionals use a valuation model. A common model is two and a half times the gross sales of the previous year. Another is the average of actual sales three years in a row, times three. There are certainly more complicated models if you ever decide to sell the business to someone else. No matter what valuation model you use, your business, no matter how small, truly has value based on its sales—and those sales come from your customers!

## Your Customer Base Is Different from Other Assets

Your customer base is a unique asset in terms of the care it needs and the maintenance work you must do. It is also different in that if you don't properly care for your customer base, you may not notice a difference immediately. It may take several months, or even longer, for you to observe and then start to suffer from a customer base that is not being cared for enough to keep them satisfied and coming back to you.

Customers may complain, like hungry cows, when they're dissatisfied. That's not so bad, because at least you know what's wrong. The big problem is when they don't complain. If your customers don't voice their dissatisfaction and you don't notice it, you'll soon be in trouble. If that problem persists, you'll lose those customers, and winning them back will be a major project—if you ever can.

This is how all wine begins.

A customer base also differs from other business assets in that your work to keep those customers happy requires less physical labor and more tactical effort and marketing savvy. You will run into that right brain–left brain kind of combination all the time in small rural enterprises. So follow along as I offer my best tips to keep customers happy, and learn from Renee and Bob Guilliams of Raven's Glenn Winery.

> You've got to pay attention to how your customers are feeling and ask them questions about their satisfaction level. **They may complain about a big problem, but if there are little things they don't like about your business, they may never mention them.** They'll simply go elsewhere.

## How Do You Protect Your Customer Assets?

Start working on customer retention as soon as you get your very first customer. So many businesses frame the first dollar they make and hang it on the wall, and doing that is more symbolic than you realize. Sure, you've got a little money, but now you've got an entirely bigger job than you did when you were getting ready to open. Now you've got a customer—represented by that dollar—and, hopefully, many more. You've got to pay some attention.

You'll likely be super-motivated when you start your food or farm-based business, and that enthusiasm will carry you a ways. In fact, it may be contagious enough to secure those first few customers. But maybe not. Many, many small farm start-ups lose more customers in the first year than they gain! The scenario is common because you're so hyper-busy in the start-up phase that you're focusing on, well, everything, and it's hard to pinpoint which task is essential and which task is just noise in the back of your mind.

Customers for foodie and farm businesses are usually generous and excited about new options. They will share your enthusiasm for a while and will put up with your mistakes and your learning curve, too. But eventually, they'll crave consistency and service, and they'll need to be reminded to buy from time to time.

So what do you do to protect this customer asset? You've got to build a customer database, packaging customers together as a group so you can make taking care of and communicating with customers a manageable task.

Renee and Robert Guilliams, owners of Raven's Glenn Winery, at the tasting bar.

## Meet the Gang at Raven's Glenn Winery

Bob and Renee Guilliams are not new entrepreneurs. In fact, they've been in business together since the 1970s, when they bought a long-term-care facility in Ohio, which they ran and grew through 2009. Under their ownership, it billowed to 320 employees. "By the late 1990s, we decided that the business had matured and that we were really burning the candle at both ends," Bob recalls. "Still, we had a lot of energy and wanted to do something different for our 'retirement' years."

Bob's initial plan was that the couple become skiers, though Renee wasn't keen on the notion at first. Eventually, he coaxed Renee out to Lake Tahoe to learn to ski, and he says she caught on rather quickly. But skiing was not to be the retirement avocation of the Guilliamses.

On one of those visits, the trip went far afield—in a good way—when Bob and Renee ventured westward to the Sonoma and Napa Valleys in California. The trip sparked an idea that the couple couldn't shake—starting a vineyard. "It was one of those things that you just sit around the kitchen table and think about. We decided it might be something we could do," Bob says with a lilt to his otherwise steady voice. Raven's Glenn Winery was born.

They started to sell off pieces of the long-term-care facility and bought 100 acres in West Lafayette, Ohio, for a winery and vineyard. They also built a new home on the grounds. The original 2,500-case-per-year boutique winery, opened in 2003, quickly outgrew their "retirement" plans. The Guilliamses needed help and turned to their son, Beau. He joined the company and they followed a couple of his suggestions; smart move, apparently. Beau thought they would get more business if they moved the winery down to a place

## How to Create a Customer Database

Don't rely on your mind to remember all your customers' preferences and their previous purchases—that's what a database is for. The concept of a database is simple: You add each customer to a computer program and record information about their purchases and preferences that you will later use to stay in touch with them and drive more sales.

Creating a customer database is easy, though there is some labor involved in data entry. You can use the programs that come with your transaction software, or you can keep it extremely simple and collect names, addresses, and e-mails and enter them into an e-mail management program such as Microsoft Outlook to create a distribution list. That's the on busy U.S. Highway 36. They also added a restaurant the following year.

Soon, Renee emerged from her very brief retirement to become the hospitality manager, Beau took on the role of wine maker, and Bob now handles development, business, and marketing.

No one at Raven's Glenn had any experience in wine making or the restaurant business, though Bob says that Beau has become an excellent wine maker. "We thought we knew food service because of the care facility," Bob says. "We were wrong! We've had our problems, none insurmountable, but there were definitely surprises." Knowing they needed expertise, Bob hired consulting wine maker Tony Carlucci—a decision that made all the difference in terms of quality.

Wines made by Raven's Glenn appeal to a wide variety of consumers. "We knew we needed to have a kind of wine that most people in the Midwest appreciate." Raven's Glenn makes 20 varieties of wine. "Not everyone wants Bordeaux or Chardonnay. A lot of Midwesterners like a little sweeter wine, and there is a cross-section of people who like fruit wines," Bob says.

"We also needed to have destination and a reason for people to come to justify that capital investment," Bob explains. "The idea is to find a wine at the tasting bar that you like and then go into the restaurant and create a whole meal around the wine."

Wines can be bought at the winery or in the restaurant, but the majority of their business comes from selling wine at retail stores all over Ohio. For that, they use a distributor. Wines can also be shipped, but due to federal and state regulations governing alcohol sales, the Guilliamses have opted to ship only within Ohio.

way I did it for Aubrey's Natural Meats, and it worked fine because I was small enough that I didn't need sophisticated accounting or transaction software and because I sold primarily at farmers markets and didn't have computer equipment on site.

The process I used was straightforward. If a customer bought a product, even with cash, I asked them to join our e-mail distribution list. About 80 percent agreed to join, even if they were just buying one $4 package of ground beef. They wrote down their name, address, phone number, and, most important to me, their e-mail address. I later took that data and entered it into my Aubrey's Natural Meats distribution list in Outlook. Then I made a note next to each customer whom I knew something about.

Having a sign-up sheet at your farmers market booth or your retail store is simply a must to keep adding customers to the list. Train all employees to ask customers to sign up, and make sure they either enter the information themselves or read over the customer's shoulder to make sure the writing is legible! Try to watch a customer writing out their info; if you can't read it, you can't add it. Yep, I've screwed up on that one, too.

In addition to actual customers, there are a number of ways to add potential customers to your database. One easy way is to offer your customers a small discount for referring someone who actually makes a purchase. That referral then joins the customer list—even if they don't end up buying.

It may sound silly, but **don't forget to ask customers to put their names on your customer sign-up list!** Before I started to specifically remind people, I had a list of e-mail addresses but no actual names of my customers.

People have grown accustomed to entering just their e-mail address, and many personal e-mail accounts feature usernames that are, uh, not the same as the owner's given name. I mean, you're going to feel a tad silly sending an e-mail newsletter out to big daddy@aol.com and yougo girl@hotmail.com if you've forgotten to ask those customers for their actual names. Besides, it seems both impersonal and totally unprofessional.

You can also offer referrals from vendor to vendor at the farmers market. Try partnering with someone who sells an item that complements yours (for example, you sell cheese and they sell bread or wine) and ask customers to visit their booth, and vice versa.

An additional idea to acquire database members is the old "enter your card to win" drawing. You set up a bowl or box and ask customers to drop their business cards into it. The prize—which perhaps is drawn once a week or once a month—is one of your products.

Signing up customers at the point of purchase is probably the best way to build a database. The next best way is the Internet. Most web sites are designed with an e-mail capture form and an opportunity for the web site visitor to sign up to receive more information. I get quite a few new customers this way.

Ways to transform those interested people into customers include instant contact (immediately e-mail back to thank them for joining and confirm their addition to the e-mail list) and added value (offer them a discount on their first purchase in exchange for signing up). Keep in touch with the people in your database at regular intervals so that newly added members get something interesting from you relatively soon after they join.

The only caveat for gathering customers for your database is not to bother adding people just to add people. I add people if they buy something, of course, and also if they seem really interested and want to learn more before they buy. Also, when you're collecting names in person, don't force people to join if they seem apprehensive about it; give people the option and then back off. If you are sincerely using the list for your own use only and not selling the names or sharing them with other organizations, be sure to let your prospective customer list members know that, too.

Beau Guilliams has become an excellent wine maker. Here, he tops up the aging barrels.

## Organizing Your Customer Database

Once you've begun collecting customers and getting serious about creating a customer database, you'll need to organize your data. How you choose to organize it certainly depends on you and should be based not so much on how to sort the customers but on the easiest, most obvious ways you can use the list. Alphabetical order is obvious, and most programs default to it, but depending on the programs you use, you can create subcategories that fit certain criteria.

You can also organize names by creating several lists that are divided into workable groups for your needs. For example, you might organize by farmers market if you serve several, making a separate list for the customers who shop at each market. That way, if you're going to miss one market on a particular day, you can simply notify that group and not send a blanket e-mail that goes to people who don't even live in that town. You can organize by region if you travel to certain areas, by event if you work a lot of annual festivals and fairs, and by year if you'd like to know when you added people and how your business is growing.

**Before adding customers to your mailing list, ask their permission** or make sure it's obvious from what you say. For example, if you have a sign-up sheet, make sure it says something like, "Sign up to receive our e-mail communications." On your web site, be sure the contact form says something along the lines of, "By signing up, you'll receive periodic information from us."

When you do send out e-mails, give people the option to unsubscribe somewhere in the e-mail, and make that process as simple as putting "unsubscribe" in the subject line and shooting an e-mail back to you. But also point out that the customer is welcome to forward the e-mail to a friend or colleague.

An organization tactic that worked for me was categorizing people by product type. Because I sold both beef and pork, it was good to have an idea of what was selling by product type and to whom. You can also group people together in a list based on a purchase price threshold. Again, this strategy helped me categorize and market to buyers of half beef and pork differently than I marketed to buyers who only bought one package at a time. You can always group your smaller lists back together for quick communication to everyone.

After you've collected, entered, and organized the names, your next step is a bit more record keeping. Make notes about what customers have purchased and also any problems or preferences they may have. You can put these notes on the customer invoices that you store in your accounting software. You can also add notes in your e-mail address book, along with the customers' contact information. Your best bet is to refer to these notes before contacting a customer so you will not only be prepared, but also demonstrate your sincere knowledge of his or her needs when you speak.

## Storing Your Customer Database

This should go without saying, but I've known small-farm entrepreneurs who've met with major disappointment, so it bears mentioning. Save and store your customer database! After you've spent time securing customers and building a database, you sure don't want to lose it. You'll never replace even 40 percent of the database if it is lost, so take great care with it.

Storage can vary depending on your needs and your techie ability. At the very minimum, keep the database in a word-processing file or in your e-mail distribution list and print a copy periodically. Better, and in addition, get an external hard drive and save the list on that hard drive after each new

## The Raven's Glenn Customer List

Bob tries to add most visitors who come to the winery to his database. Still, he believes the opportunity to add customers is a privilege. "We are really sensitive about requesting information from people," he says. "We promise never to give out their name or telephone number. It still surprises me how many people will sign up in our guest book."

He primarily asks people for their e-mail address and zip code so he can track where the wine is going during the busy season. This information helps him make decisions about what retail stores he would like to carry his wine.

Six times a year, he sends a customer newsletter by e-mail—a schedule he deems often enough to inform but not too frequent. "We don't pester people. If there is something legitimate, not just hype, we send it out," Bob says. Most of his newsletters have information about new releases and special events at the winery. (You can see an example of a Raven's Glenn newsletter in the appendix of this book.) "If you abuse a customer with junk, they'll treat your news like junk," Bob says.

Collecting customer names in a database is useful for sales, but tracking the exact number of sales that result from e-mails is difficult. "It's hard to know if a customer came in because of a newsletter, because so many come into the winery over and over to buy wine and shop," Bob says.

Although he has considered starting a wine club—a common customer retention idea used by wineries all over the world—Bob hasn't taken the plunge because alcohol sales regulations in Ohio prohibit him from giving more than a 10 percent discount, and that discount has to be on a case or more. "Until the regulations change, it's just not something we'll do because I don't think there is much incentive," Bob says.

addition. Burning the list to a CD or adding it to a memory stick that you store in a safe is also a great idea; that's actually what I do with mine.

No matter what you do, keep your list protected in several forms. You don't want to lose it if your computer crashes!

While you're taking time to save, also consider doing some customer database clean-up. Again, updating the list as you go keeps the job manageable and increases the likelihood that you'll do it. Updates include immediately removing those who ask to unsubscribe, deleting e-mail addresses that no longer work, and removing customers who have moved or won't, for whatever reason, be able to buy anymore.

**Start entering names and organizing your customer database as soon as you get started.** You don't want to have hundreds of customers to enter in a huge stack; you'll probably never get to it! Stay on top of the organization or ask someone else in the family to do that task at least once a week.

## Using Your New Customer Database

Build it, organize it, save it—now use it! Working with your customer database regularly is the entire reason for its creation. I'll cover customer retention in the next section, but retaining customers is only one use for this database you're starting to build. You'll also want to use this list often to add value of all kinds for your customers, including communication about many things that are important to them, such as sales, promotions, and times when you'll be closed for business.

You can also use it to promote your special causes, other partner farms you support, and pet projects that are meaningful to you. Another great use is to provide useful information on topics related to small farms and food business. It could provide a venue that your customers know will always give them news and insights they didn't have before. Links to other sites and articles of interest is another great use of your customer database.

Reminding your customers that you are out there is the single biggest use of your customer database. If this sounds mundane, trust me, it isn't. Consider just how busy you are. Consider your many good intentions that go undone and forgotten. Consider your financial needs and priorities and unexpected expenses. Then consider something you really want, really like, and really hadn't prioritized, but suddenly you're reminded of it at just the right time. That is the beauty of a customer database. Not all customers will need what you're offering all the time, but those who do need it also need to be reminded that you are out there and interested in them. Communication is essential, and your customer database gives you the tool.

## Retaining Customers

Since I've beat into your head the importance of retaining customers, now is the time to share some ideas about how to do it using the customer database you're assembling. To use an old financial planning term, customer retention is like having an annuity for your business. An annuity is something you own that has cash value and also provides the opportunity to earn interest. You can continue to earn interest on it (that's like sales for your business), but you can also draw on the money in the annuity (that's like the database you've built). As long as you don't spend faster than you earn interest, your annuity can last you a long time.

Customer retention is certainly about reoccurring sales. But even if some customers are not regular buyers, they are still valuable to you in other ways. For example, they may not need your product all the time, but if they love it, they'll forward your information to others in their network, giving you access to more and more people with the same amount of work. Besides, you never know when some communication with a current customer will strike their fancy and inspire them to buy from you. In my business, for example, I have many once-a-year buyers. If I don't contact them via e-mail and remind them that they last bought around a year ago, they just might not get around to it—or they just get a message from someone else and decide not to shop with me.

**How much contact is too much?** The frequency with which you contact customers is totally individual. My best advice is to first contact them when it's important—such as what dates you are opening for the season—and move up from there. For example, I had a seasonal business with Aubrey's Natural Meats, running from May to October at farmers markets, but I accepted orders for large quantities year-round. I contacted my customers weekly with specials in the summer and then once a month in the winter. The winter contact was like a reminder; I just wanted them to remember me when spring came—and if someone wanted a half beef in the meantime, excellent.

You'll learn as you go how often the appropriate contact is. The main thing is to be interesting and add value; don't get annoying! If you've already sent a message this week, unless it's an emergency, don't bother the customer again just because you forgot to mention a small item.

This section includes a number of ideas that I have used over the years to retain customers and encourage sales. Remember, customer retention is as much about keeping customers happy and showing them appreciation as it is about getting new sales from existing customers. There are no rules for using these ideas. More than likely you'll need to implement a combination of several customer retention ideas each year.

## Coupons

Many customers appreciate coupons because no matter how affluent they are, everyone likes a deal! We've all clipped and redeemed them at one time or another, so why not use this time-honored technique in your own local farm and food business? Design coupons around a certain theme or around items that are either high-margin for you or that you need to sell. Make sure your coupon is clearly marked with how many times it can be used, an expiration date, and any other parameters you want to establish.

## Discounts

Discounts for repeat customers work wonders for sales. Many customers are used to getting a deal for becoming a new customer, but I don't like that method. I'd rather tell a new customer that once they buy, they will qualify for the repeat customer discount on their next sale. This way, you've already got the customer thinking about buying from you again.

Discounts can also apply for all customers at a certain price point. For example, we offered 40 percent off the retail price for customers who bought forty pounds or more of ground meats. This really helped move what had originally been items I had extra of. The program was so successful that we actually ended up with a waiting list for ground beef!

## Preseason Orders

Another way to keep existing customers is to offer them deals before the rest of the public. I've often done preseason discounts for those who order in advance of the first farmers market and then take delivery on that first May Saturday morning. You can also offer current customers last year's pricing for their first order—something a walk-up customer won't be able to get. My advice is to get at least half of the cash in advance for preseason orders so you know customers are serious about buying and so you're not long on product without guaranteed sales.

## Free Gifts

If you sell a variety of items or decide to partner with another small farm vendor, offering a free gift to reoccurring customers can be a nice idea. Offer the gift with their next purchase or with their first purchase of the season, or give out gifts at the last market of the year. The gift can be of low or high value. Just make sure that when they see it or use it, they think of you!

## The Personal Touch Creates Customer Loyalty

One of Bob's best tactics to win and retain customers at Raven's Glenn is the simplest—he gives them a moment of his time. "I am amazed at the number of people who get home and write us an e-mail to say they liked the wines and the restaurant. Either my wife or I personally sit down and write them back within 48 hours and thank them for being here," he says, emphasizing that returning correspondence quickly is key. This response builds loyalty because when people think they were treated well, they tell others. "They may not make it back this year, but they will tell someone else. People like to be able to talk to the owners."

Bob also wants the customers to feel special at the winery and see that the staff is professional. All staff members wear logo shirts, guests are welcomed upon arrival, and if they have questions, Bob tries to answer them. Being courteous is "so easy, and many places just don't do it," he says.

Another way Bob fosters long-term customers is by making sure his wine is widely available in large retail stores. "The idea is to find a wine you like here and be able to get it at home," he says. "That is part of building that customer loyalty."

## Publishing Photos

Agritourism ventures offer some of the nicest opportunities to encourage customers to come back at least once a year, because they set a mood and provide a wonderful experience. Bring that spirit to any small farm and food business by taking photos of customers and their families enjoying your products at your location.

Publish the photos you take of customers on your web site or send them out to the whole list in a newsletter. The idea is that the current customer feels a bit famous and new customers see others enjoying your farm. And here's another neat idea: Send the photos to the customers as gifts, or maybe even create refrigerator magnets that you can give away.

## Open Houses and Special Events

Another excellent customer retention tool is hosting special events or an open house at your farm or at other venues. Customers receive an invitation and you offer them perhaps a combination of discounts, members-only specials, or preseason prices. Or maybe your open house is a Christmas

The patio at Raven's Glenn Winery overlooks the beautiful Tuscarawas River, making it an excellent venue for parties and special events.

party thanking them for their annual business. The ideas for special events are limitless; the key is that they should reflect your business and your farm's niche.

## Members-Only Concepts

Remember the feeling of being "in the clique" in junior high school (or out of it!)? It was all about membership, and of course, membership had its privileges. Guess what? It still does! No, you don't need to re-create the who's in–who's out atmosphere. Just offer your customers the chance to have their own club with your farm as the center.

Wine clubs do a great job of this by offering members-only deals, invitations, discounts, and access to favorite items before the general public. This membership concept can take place on the farm, too. Build a Farm to Table club, set the membership benefits, and enroll your customers.

## Reoccurring Credit Card Orders

Another customer retention idea is securing their business in advance. The easiest way to guarantee repeat orders is with a credit card that is automatically billed at set intervals. Make sure the customer knows they will be billed (and receive product) at these set intervals. Create an order

## Tours and Special Events

Raven's Glenn is right next door to a golf course and just above the banks of the Tuscarawas River. The deck overlooks the river and provides pleasant outdoor seating. "The setting here is built around the destination idea," Bob explains.

The restaurant and grounds have become the destination for numerous banquets and weddings each year. Bob says they also offer fantastic tours, many of which he conducts himself. "The tours help build customer loyalty," Bob explains. He wants to offer people the chance to learn something about wine that they can use when they leave, such as food and wine pairing tips and how to approve a bottle of wine in a restaurant. "We make a fun thing out of it and people have a good time. It instantly creates that bond."

Of course, after the tour it is time to taste the wine. Bob says he can't give away free samples due to state regulations, so he charges $2. For that price, the visitor receives not only tastes, but also a logo glass. The first year they gave out 4,000 glasses; now they hand out more than 25,000 a year. "People come back and say, 'I'm here to add to my collection of Raven's Glenn wine glasses,'" Bob laughs.

agreement that clearly spells out the terms and have each customer sign it. Give a copy to the customer and retain a copy for your records.

# Using Customer Feedback for Promotion

Another great way to retain customers is to gather their feedback. The simplest and cheapest way to gather feedback is just to ask for it. Ask your customers how they feel and what they like and don't like about your products, services, and distribution options. Ask regularly and thank regularly—these are the basic customer retention principles of any small business!

Beyond regular discussions with your customers about their preferences, you can get more sophisticated and send out surveys. I surveyed my customer base for Aubrey's Natural Meats every spring and fall, trying to find out what their favorite products were, when and where they preferred to buy, and what they wanted to see improved. If you encourage customer feedback, you can use it to keep your customers happy.

Customer feedback is also a great way to improve product promotion. For example, if you sell foods, consider collecting customer recipes that feature the items you sell. Then (with permission) publish those recipes on your web site, crediting the individual customer, or make a brochure or a small recipe book. If you create a cookbook, you can give copies away or sell them at a reasonable price, donating some of the proceeds to a local group that you support. This way, you gain feedback about how your products are being used, have a promotional opportunity, and contribute to customer retention because customers get to see their names in print.

If you don't sell foods, what about creating a compilation of the best or most creative ways that customers have used your products? For example, if you sell natural cleaning products, create a brochure or even a web site with great ways people can use them, or list the commercial products they replace. If you sell medicinal herbs, find out how customers have improved their lives by using those herbs, and make a list or a little booklet of tips to give out or sell.

Be sure to ask (and get an e-mail okay so you have a record of their approval) before each use of a customer's name in any publicity efforts.

# Customer Communications Tools

Most customer retention activities require you to communicate with your customers. So do most marketing and sales strategies. The value of regular communication cannot be overlooked; this is why you spent time creating a customer database. Sometimes you need to communicate with your customers to tell them something important about your business, and sometimes it's just to keep in touch. In doing so, you want to add value to their lives. Below are a couple of my favorite customer communications tools.

## Newsletters

Offering a customer newsletter is one of the best ways to stay in touch with your customer database and remind them that you're interested in their continued business. As a consumer, you probably subscribe to several e-mail or even hard-copy newsletters. As you think about creating your own newsletter for your hobby farm, ask yourself a few questions about the newsletters you already receive.

- What about the newsletter you receive intrigues you?
- What sections do you read?
- What sections do you skim or even skip?
- Why did you subscribe in the first place?

## Quality Comes First

Although Bob is an affable and polite businessman, there is one thing he gets a bit (unapologetically) testy about—quality. "Quality is foremost. I knew from the beginning that I wanted to treat this as a business and not as a hobby," he says. He understands that not everyone wants a second career, but he believes that if you get into the business, even just for fun, you must have a respect for the industry and be committed to maintaining high standards.

"We see a lot of people in the winery business right now who think that if they make home wine, they can become a winery. If they are not properly capitalized, or whatever, then they may not evolve into something that is good for the industry." he explains. "Many people believe they can turn an avocation into a vocation," he says, but Bob points out that something as regulated and expensive as making wine is not to be taken lightly. "Make sure you take the time to research; we did research for five years before we got started."

It is important to Bob that his industry be promoted with quality products and in a positive light. "You have an obligation to other people who have done their homework and have a good-quality product," he says.

There is a difference between a hobby and a business, and that difference is certainly obvious when it comes to food. "This is a complete and total reversal from our other life," Bob says with a laugh that belies the truth of the statement.

- What makes you continue to receive the newsletter rather than hitting "unsubscribe" and removing yourself from the list?
- What things do you wish were in the newsletter that are not?
- Have you ever mentioned this desire to the newsletter sender?

The newsletters you get can provide great ideas (and some things to avoid). So from now on, examine every newsletter you receive for design, content, and general value. Then gather the ideas you like and package them into your own newsletter format. (The appendix of this book includes an e-mail newsletter from Raven's Glenn.)

Making a newsletter can be as simple as typing up a document and either printing and mailing it (sort of like the annual Christmas letters your Aunt Agnes sends every year) or copying and pasting it to your e-mail

When contacting customers as a group, **don't forget to use your e-mail program's blind cc function.** This means that you do not show everyone's name and address to everyone else. Respect people's right to keep their e-mail addresses private!

program and sending it out that way. Or you can type right into the body of an e-mail and send it away.

There are online newsletter management companies you can use for a relatively low price. Constant Contact is one choice. You provide the company with your information and your customer database (they can collect new sign-ups for you as well), and use one of their many templates to design your newsletter; the company sends the newsletter at your preferred time. You can also find free newsletter templates online or buy a CD from an office or electronics store that comes with many options for newsletters and brochures. Microsoft Word has a built-in basic newsletter template, too.

No matter what format you choose, keep an archive of your newsletters in hard copy or digitally so you can refer to them from year to year.

### What to Include in a Newsletter

Your newsletter can include just about anything that is important to you and generally relevant to your customers' interests. While many customers love the personal aspect of knowing their small farmer, walk the line, as they say, between telling your personal story and boring people senseless with your own news and nonsense. Remember, the newsletter doesn't do its job unless it adds value. So make your information seasonal and different every time. Talk about new products, describe the happenings on the farm, include a recipe or two, and offer discounts or incentives.

Finally, in each newsletter, always include your contact information in a convenient location.

Make sure your newsletters are regular and that you have enough pertinent information to fill them. Send out newsletters only in months or seasons where you are apt to have news for customers, such as opening for the season or new products or pricing. The same goes for the end of the year, changing or discontinuing items, and changes to selling venues. Make it just one page if that is all the space you need to inform your readers.

### Does Design Matter?

Design and layout don't matter a great deal, but many design experts will tell you that if you create an attractive template and use it over and over again, the consistency will be attractive to readers. The most important aspects of design are to make sure the layout is user-friendly, easy to open with a variety of Internet and e-mail programs, and consistent. The layout should be attractive, but not so busy that it is off-putting or confusing. Basically, let the design show your creativity and showcase the uniqueness of your farm, but don't feel like the newsletter has to be an artistic masterpiece.

## Press Releases

Another communications tool you can count on to help you promote yourself is the media. In chapter 9 I mentioned free editorial articles and being interviewed. The way the local media finds out about you is through a press release that you send to them.

With a press release, you write a short article from the point of view of a reporter and submit it to media outlets. For example, as the reporter, you would actually quote yourself or another person in your business or family. You write the story in a format reporters call the inverted pyramid—that is, most important news first, least important toward the end. You also provide a contact person for more information. Here's an example:

> *Aubrey's Natural Meats, LLC, owner Sarah Aubrey*
> *announces a new farmers market location for 2009.*
> *"We're so excited about setting up a booth in this new*
> *location. Our customers have requested this for years and*
> *we are finally able to serve them," said Aubrey.*

Make sure to carefully proofread anything you send out so there are no typos, misspelling, or grammatical errors. Don't reply on your computer's spell-checking function to catch everything for you; have several people look it over.

The local press may decide to publish your press release just as you've written it, or they may decide to interview you themselves, based on the information you sent. Of course, unlike an article written about you where the author usually knows in advance that it will go to press, you can't count on all press releases being published because you are soliciting the media outlet, not the other way around. Still, if you send press releases to media with which you have a relationship, such as the communications director of your farmers market or the local rural development office, you'll be more likely to get your news disseminated. Also, build a relationship with local newspapers and even small business and trade magazines.

Renee keeps things clean in the tank room.

In this book's Sample Business Documents, you'll find a sample press release that can be modified to fit your media communications needs.

## Handling Customer Problems

The final key point to customer retention is dealing with customers when concerns or problems arise. Don't assume that just because a customer has a complaint about your products or services that you've already lost their business. Even in potentially disastrous situations, you can often salvage the customer's business and even create lasting loyalty just by handling the situation with genuine care.

It has often been said that how you react under pressure says a lot about you personally. This statement is never truer than when dealing with disappointed customers. Courtesy, respect, and overall kindness, spiced with empathy and understanding, will go a long way toward winning customers over and turning a frown upside down.

## Customer Returns, Exchanges, and Refunds

The first customer care issue is returns. This situation can be especially tricky for food businesses because the items you sell are perishable and require proper care and handling. You have no way of knowing how the customer stored the food after it was sold to them. Offering returns and exchanges can also be a problem when the faulty product no longer exists because the customer ate it.

My advice is to draft a policy on exchanges and refunds and post it in a conspicuous place. Explain your returns policy in person to customers who seem hesitant to buy or appear unwilling to take safety precautions. (I once refused to sell meat to a customer who had biked 45 minutes to the farmers market in July with no cooler. He planned to shop around town a while longer, with the meat *in his backpack,* and then bike home! Can we say food poisoning?)

Some producers believe they should allow returns anytime with no questions asked. I, however, do not. I've seen return policies on food and wine that include asking the customer to call the moment they notice a problem, not accepting returns if the customer doesn't bring back the unsavory food or what is left in the package, and not accepting returns or exchanges on food after a certain number of days.

Overall, your returns and exchange policy will need to make sense for your particular product type. Ask other vendors and find out what is common in your area.

## Customer Complaints

Along with returns come complaints. To me, complaints are more potentially detrimental to your business than returns or exchanges because some complainers just want to be heard or enjoy making a scene. Many complainers also want to prove publicly that they are right.

I've found that most consumers who come to you with a need to return something do it for an honest reason. Most of the time, they feel a bit guilty and don't want to inconvenience you or cause your farm financial hardship. Complainers, I'm sorry to say, are often different. I've had my share of out-and-out complainers—every business gets them—but I am happy to say that it is not a frequent occurrence. Many people who buy from foodie and farm vendors do so out of a love for the products offered and for the enjoyment of meeting people like you. Most of these customers will never complain.

You've got to handle complainers with kid gloves, and you simply can't ignore them. Return the call or e-mail immediately, and work as

## Complaints, Concerns, and Returns

Despite the great service, food, and wine at Raven's Glenn, there are always problems, though mostly they are small ones. Some are as simple as replacing a bottle of wine that was damaged or sending back a meal that wasn't properly cooked. But no matter how trivial, Bob says customer service is essential, especially if the customer is dissatisfied.

"Don't ever argue with a customer," Bob begins. "And never, ever, do the old, 'Well, so and so's off today and it's not my problem' either. Most of the time all they want is to be heard. So keep your mouth shut and customers will tell you what is going on."

Bob says problems should be dealt with immediately and that the business owner should take personal responsibility for fixing the situation. "If you can, fix it, if doing so is within your limitations. If not, be honest about it," Bob suggests, adding that sometimes a problem can't be rectified to a customer's liking or that he is limited by alcohol sales regulations. "Around here we go overboard to fix things. Admit there is a problem and ask what you can do for them."

Handling returns with food and perishable items can be tricky. Sometimes it is also difficult to convince customers that harmless crystals and the like are not flaws in the wine. For example, Bob says that if wine is too cold for a prolonged period it will be "shocked" and little crystals of potassium tartrate (present in all wines) will appear at the bottom of the bottle. Bob knows that this is caused by poor handling of the bottle, but convincing the customer who wants a return can be difficult. His advice: Just take care of it, but be sure to explain to the customer how the problem was caused and how to prevent it in the future.

"The bottom line is, don't overlook customer service and customer satisfaction."

patiently as you can to rectify the situation. Some complainers actually want something, such as a refund, while others just want you to hear them complain. It is your choice what you do with complainers, but again, don't ignore them; handle their problem personally and as quickly as possible.

There are many options for dealing with complaints. You can offer a refund or exchange, per your policy, or you can simply offer an apology and an ear to chew on. Another idea is to give the complainer an added incentive or a free gift or a coupon—along with that ear and apology, of course.

# Managing Customer Expectations

When it comes to refunds, exchanges, and complaints, you will probably have to handle each situation individually. But don't set a precedent that you're not prepared to follow time and again. Oh, did I ever make this mistake when I was first in business! Here's a little story for you.

A woman who had never bought meat from a small farmer approached me at a farmers market and wanted me to convince her that she should buy a quarter beef (I had a sign out advertising quarters and halves). I responded pleasantly that I'd be happy to talk with her in detail about the quarter beef and answer any questions, but I wasn't going to convince her because it was her decision in the end. She scowled and retorted that if I wasn't a better saleswoman than that, maybe she ought to go to a competitor.

I fought the urge to invite her, not at all politely, to do so immediately, and ended up asking her to sit down and visit with me. She did, and she bought the quarter. I was really glad for the business, but felt uneasy about the bossy woman and her poor attitude. Later that week, a friend of hers called and also bought a quarter. I thought I had done a rather fine job of saleswomanship after all!

When delivery day arrived, the customer hated everything, though I felt I had held up my end of the deal. She said the packages were too plain and she'd hoped for more attractive labels (at that time, I reserved my "fancy" labels for farmers market sales and used the less expensive labels for bulk orders). She said the sizes of packages didn't all seem perfect. I explained that since she'd asked for cuts of different weights there was some variation. And even though I personally carried the boxes down to her basement and put them away in her freezer, she didn't like the way I organized them. Aggrr! I left convinced that would be the last time I dealt with her. Oh, but it wasn't!

That night she called back. The meat had a slightly off smell, she said, and she just didn't like the color. And besides that, the ground beef seemed a tad too lean. I felt like screaming by now. I tried to pacify her and explain away these problems, but there was no doing it. The woman got nasty. Finally, afraid that she might write a scathing letter to the editor of the local newspaper, I agreed to come get the product and refund her money. Big mistake.

The friend of hers who also bought from us (my husband had delivered to the friend without incident) called the next day saying she'd heard about the *beef recall* and wanted her refund, too! I thought I might faint from frustration, and then pure white-hot anger seethed through my veins. First of all, the woman was a menace. Now it seems she had become a liar, too.

But I'd set a precedent with the first customer by taking all her meat back, and now I was stuck. I refunded the friend and picked up the meat. My husband and I ate really well for a while—after all, we had an unexpected half beef in our freezer (which tasted perfectly fine, of course). But our budget was tight that month because of the loss of two big sales.

A couple of weeks later, I saw the snippy customer from afar at the farmers market. Fortunately, she never came by and I was glad. But she was still lingering. A young couple approached me and the wife said, "We just heard from this lady that we can buy a quarter beef and if we don't like it you'll come pick it up and refund all our money, risk free!" As you can imagine, my face blanched white even though my cheeks were rosy from the hot July day. I simply said, "I have no idea what you're talking about."

## So How Do You Keep the Farm?

The moral of this story is, don't lose your head when you are negotiating with complainers. The best thing to do is decide if the person is a complainer (as the customer in this story was) or has a legitimate need for help and resolution from you (truly, most of your customers). Sort out the difference if you can and deal with them as objectively as possible.

If you're going to offer a refund or discount, I don't think you should ever offer more than the original price the customer paid. After all, you started this enterprise for yourself and your family, and a person with an unmerited or overblown complaint doesn't deserve a share of the farm! How much you follow the old adage "The customer is always right" depends on you. My advice is to be fair, be firm, and be friendly. Good customers will respect you for that!

---

### Bob's Best Practices

- Treat customers the way you'd like to be treated. A personal touch is best.
- Hire consultants where necessary to increase your potential for success.
- Take time to do your research before you get started.
- Sample food products similar to yours from all over to compare and get ideas.
- Keep the quality high regardless of whether you are operating a hobby farm or starting a serious business.

# Part Four

# The First Year of Your Journey

# Chapter 11

# The Lifestyle Change and First-Year Transition Tips

### Learning Objectives

&#10086; Consider various lifestyle changes that will result from starting a new venture.

&#10086; Consider other people's reactions to the changes you make and how they will affect you.

&#10086; Realize that your new awareness of food and farming comes with responsibilities.

&#10086; Learn the six factors that are most important for surviving the first year.

&#10086; Gain tips to deal with the first year of your new business.

&#10086; Meet Jenna Woginrich, part-time small farmer, blogger, and mountain musician, and full-time graphic designer.

## *Of Rural Living, Food, and Farm*

Here we are at the final chapter of this book about your journey from consumer to producer. We've reviewed, evaluated, pondered, and planned. I've inundated you with figures and challenged you with ideas. This is a business book written from the point of view of experience, but it's not intended to be a textbook. I want you to learn while you enjoy, and I hope you've been able to do that.

This final chapter is meant to be more approachable than some of the early ones. Take a breath. This chapter is like turning to the lifestyle section of the Sunday paper; the hard business topics are now read, and you're looking to be intrigued and entertained, and maybe pick up a few last tips before starting your week. So curl up with a cup of tea or a glass of cognac and relax into some lighthearted advice from my voice of experience—and even a little joking around.

This chapter is divided into two parts. First, I'll cover some general lifestyle changes you may experience and my thoughts on dealing with those changes. In the second part, I'll share with you my best first-year transition tips for any new food or agriculture business or hobby.

You may have seen the 2003 movie *Lost in Translation,* in which Scarlett Johansson and Bill Murray played two Americans trying to find their way in Japan. Through a series of unexpected events, the characters learned a lot about themselves, but it wasn't easy for most of the journey. The hopeful purpose of this last chapter is to help you avoid becoming "lost in transition"—along the way from digging in the dirt and transforming it into something on which to dine.

In any small business, the adage "What you don't know can't hurt you" is, quite frankly, a lie. I wish I'd known more about the difficulties of the first year before I started out. But I made it through. There will be some challenges for you too, and I'll cover a few here.

Before we go too much further, I'd like to tell you that I'm so excited for you! I'm also just a teensy bit jealous. Launching a new enterprise is exciting. If you've gotten the entrepreneurial urge, you'll never feel more motivated than in the first few months. One of the keys to surviving the first year is balancing that enthusiasm with work. That also includes harnessing your enthusiasm to keep you going when you face challenges and tempering it when you're inclined to make a rash choice "just for fun." The entrepreneurial rush is exhilarating; your job is to enjoy the ride, but make sure you can afford the car!

## Lifestyle Adjustment

No matter whether you're considering a move to the country or you've already made the leap, or if you're planning simply to become an urban gardener or a small-town farmer starting a CSA, this book is about lifestyle change. All the readers of this book have something in common. It might be an interest in less consumerism, less commercialism, and finding food made with care, flavor, and dedication. It doesn't matter what decisions you're making as part of this lifestyle change and business start-up opportunity, or even if you've picked up this book for entertainment and education about what's out there in the realm of hobby farms and local foods; your lifestyle can and will change as a result of the interests you have in

## Meet Jenna Woginrich, Twenty-Something and Longing to Farm

How does a 26-year-old small-town girl from Pennsylvania end up on six acres in rural Vermont? Well, it could take a book to explain—and that's exactly what Jenna Woginrich decided to do in her debut work, *Made from Scratch: Discovering the Pleasures of a Handmade Life*. Jenna, who works as the web designer for a fly fishing company, now lives in the Green Mountains of Vermont near the town of Sandgate, but she didn't grow up in the country.

In the small town of Palmerton, Pennsylvania, population 5,000, Jenna grew up the daughter of a businessman and a teacher. Her childhood was basically middle class, and though she lived in a small community, she says she was always "basically a town kid" and that even in small towns there is a difference between being a farmer and being a consumer.

Growing up, Jenna wasn't sure what she wanted to focus on. Eventually, she decided to pursue a career in graphic design. She did not move to a big Eastern urban center as many of her friends did, but rather headed to the South and settled in Tennessee—still in town—taking her first job there as a web designer. It was in Tennessee that her love of rural living began to grow. "I just fell in love with Appalachia," she recalls fondly. "I love the history there,

this area and the growing opportunities that exist to quite literally cultivate those passions.

Are you excited about this shift in focus and thinking, or are you apprehensive? Are you making a big change or simply taking some time to learn about other ways of living? In this chapter you'll have the opportunity to read some of my ideas about making the lifestyle transition easier, but just as important, you'll meet a young professional who has decided that being self-sufficient and living a life rich in labors of love isn't something from a bygone era. Change is hard, but it always results in a new outlook on life.

## Lifestyle Changes to Ponder

There are a host of lifestyle differences between country life and city or town life. The little fable about the country mouse and the city mouse is still accurate in our modern world. Many differences are subtle, and some old cabins and historic settlements, and I love the rural feel of the lifestyle."

As her desire to move out to the country grew, Jenna experienced a life-altering moment—she thought she was going to die. She and a friend went hiking in the mountain woods one afternoon and came across a river with a large set of falls. Some young people were jumping off the falls into the rushing water below, and Jenna and her pal decided to give it a try. Jumping off, Jenna came within inches of hitting her head on the side of the cliff ledge. Instead of feeling exhilarated after the close call, she couldn't shake the feeling that she was missing out on something. "After I nearly died, I just kept thinking, 'Why am I doing this here?' So, I decided to find a job more rurally," Jenna says.

She began looking for a job in an area where she could live on a ranch or farm. As it turns out, she landed in Idaho. The major move didn't daunt her. "I think if you want it, do it. You just have to do it," she says emphatically.

Oh, and about the jumping incident that scared her into jumping off into rural living: Jenna did go back to the area a year later and spent some time sitting under the falls. "But I didn't jump again!" she says.

are quite obvious. I've always been amazed at the things my town friends take for granted when it comes to conveniences and delivery services. But there are other factors, too. They may make your life a little different at first, but a new lifestyle is what you're looking for, right?

## Fewer Conveniences

Convenience may be the single largest difference in today's city versus country life. When you don't live in town, you just don't have access to some of the things you once enjoyed. Convenience is not just about restaurants and gourmet shops, either. Consider that in the country there is less access or no access to cable television or high-speed Internet connections. You'll have to search out alternative sources for staying connected and productive. For example, I subscribe to a radio frequency service for my Internet that actually works quite well. Others I know use satellite Internet.

Many rural areas don't offer trash removal service, and rural businesses may not make house calls for services or deliveries. Pizza delivery is one

Living the rural life definitely takes some getting used to.

thing a friend who moved to the country mentioned that she missed the most! In my household, we once did convince the local Papa John's to deliver to us. Oh my, was that funny—and inconvenient. Here's what happened.

There must have been a new manager on duty at the Papa John's, because when we jokingly asked if they'd deliver, he said, "Sure." Surprised, we gave them our address and our order. An hour later we received a cell phone call from what sounded like an unambitious high school student of the variety who often work at local pizza chains. You know, the "Dude, where's my car?" type. Anyway, although the pizza manager had plotted the location of our house online, "Dude" was having no luck finding it and had been driving around for the better part of an hour. My husband got on the phone and tried to direct him, but after another half hour, he had no luck. Finally, exasperated, my husband told him to park the car and have a smoke (to which "Dude" replied that he needed one), and he'd come find him. It didn't take a local like Cary long to locate the delivery guy. While our pizza was cold, it was also free. We were done with delivery after that!

Besides some of the trivial conveniences lacking in the country, you also need to realize that emergency response times can be slower. For example, many local fire departments are made up entirely of volunteers. Hospitals in rural areas are smaller and not ordinarily as well equipped as urban medical centers. Have a plan ready for a medical emergency—will you drive to a larger area for emergency services, or have you investigated the local facilities? This is especially important if you have kids or are pregnant, elderly, or have a medical condition. Know the limitations of local medical care and decide on your best option.

Jenna takes a moment to get personal with a chicken.

## Greater Responsibilities

As you take on the role of producing your own food, even if it is just installing a large backyard garden, you've added a responsibility to your daily routine for a season or longer. Ask yourself what the responsibility means to you. Are you prepared for it? Have you allocated the time to learn, study, and simply do your chores?

Animals, especially, increase your responsibility similar to the workload involved with having children (though certainly not *that* much!). If you're totally new to livestock, including little livestock such as rabbits, birds, or goats, spend time learning, reading, and visiting with experienced people before you invest in animals. Care and the responsibility for their lives is something no one wants to screw

On pages 234 and 238, **you'll find a lighthearted but accurate description of country life and some of the generalities you'll want to make yourself aware of.** The two columns originally appeared in the weekly agricultural tabloid *Farmworld*. They were written by Midwestern ag radio and print media newsman Gary Truitt, owner of the magazine *Hoosier Ag Today* in Indiana. Gary's insight is amusing, but he also points out several key idiosyncrasies that, if known by new farmers and recent rural residents, will make them talk and feel like locals in no time.

up, but there is more to it than that. First of all, when you add livestock, you change the land and environment around you. You have waste to manage

If you keep any kind of animals, you suddenly have a huge responsibility. Who will take care of them when you go away for a day or two?

and feed and water resources to consider, and, if you have grazing animals, you need to be mindful of the animals' impact on the land and environment.

The other responsibility is having a plan for, or at least access to, disposal of animals. This doesn't only matter if an animal dies—which is certainly a fact of life with livestock. You also need to know what you can do with animals when you no longer need them or can no longer handle them.

Be conscious and responsible with these choices and seek advice that is specific to your situation. Talk to veterinarians and cooperative extension specialists in your area for advice as needed.

## Trial and Error, and Error, and Error

Any new venture involves a period of trial and error, of course. Reading this book, as well as other resources, is one way to help avoid some of those errors. Still, when dealing with agriculture and foods, trial and error goes up because agriculture is an inexact science. You can plant and plan, but factors beyond your control will always affect your crops and your sales. Be prepared for this fact, and be willing to accept it.

## Weather

Weather is always a factor when you live and work in the country. Today we have four-wheel-drives and weather radar, so storms aren't usually surprises. But the impact of weather, even just the regular day-to-day weather, makes life different. The forecasts might be the same, but outside the concrete canyon of town you're going to see different conditions.

It is often cooler in the country because there is less asphalt, and certainly wind is a major difference. If you live where we do, with no trees, the wind really blows when the fields are cleared each fall! Many farmers plan for these things with tree-line wind blocks and extra insulation in the house and barn.

With livestock, make sure you've got good shelter for inclement weather. With your crops, consider starting plants inside or purchasing a greenhouse shelter.

# What If You're Not Moving to the Country?

If you're starting an urban garden or just becoming more conscious about your eating and where your food is produced, your lifestyle change will certainly be less dramatic. Still, with this new emphasis, you'll experience lifestyle changes of a more philosophical kind. You may even meet negativity and mocking from friends, family, and neighbors.

For example, I have a girlfriend who lives in Indianapolis. In some ways, she'd love to move to the country, but in the end, she simply prefers her children's schools and the conveniences of town. So, to give herself a taste of country life, she is developing her own garden and learning to can and preserve. She's involving her young sons in the process and teaching them to understand food and how important it is to live sustainably.

Even though her intentions are good, she has run into some local resistance. For example, a neighbor complained that her garden looked a little, um, wild, because, the neighbor intoned, "It was growing up without looking groomed." My friend explained that while her garden had been weeded and well tended, she was not using pesticides and wanted the plants to be natural. She also mentioned that it was not a landscape, and she didn't want her squash and other viney crops to look like manicured hedges.

Another problem she ran into was that a couple of her son's friends' mothers didn't think their kids should eat her homemade canned items and would prefer that she "open something from a box." When she heard this,

## Jenna's Farming Experiment and Animal Adventures

When Jenna moved to a rural area near the town of Sandpoint, Idaho, she worked another day job as a web designer while she began to learn to farm on a small scale. "During that time, I started renting an old cattle farm with this big old house," she says. The family still used some of the buildings on her leased farm and raised hay off the pasture. Jenna got to know Dianna Carlin, who was technically her landlord but also became her small-farm mentor. "I had a garden, kept chickens and angora rabbits, and had a hive of bees," Jenna says of her first entry into rural living and livestock.

Her animal adventures and the general surprises of living in the country began on the first day. "There were just a bunch of little things I learned," she recalls. "When Dianna gave me my first chickens, I had a large box ready and lined with hay, but Dianna said they couldn't go home yet. We had to wait until dark. Dianna said that you want the birds to settle in their new roost in the dark so that they wake up in the morning in the new roost and still do their thing."

Jenna was also amazed about some of the basic biology of livestock that lifelong farmers take for granted. "You know, cows can have twins, but when the cow has a male and a female twin, the female can be sterile," she says, clearly still amazed by this notion. "This is a ridiculous fact, but it was told to me like everyone already knows except me!" she laughs.

she was stunned. Didn't these moms realize something canned fresh was *far* healthier than packaged food? She felt the need to educate them and share with them the information she had found out, as well as how much her kids liked being part of the process. Some moms became believers. Others said their children weren't eating at her house anymore.

I've heard similar stories from friends who've been learning about herbs and natural remedies. One friend's mother worried she was so "into it" that she feared her daughter was dabbling in witchcraft! These examples seem ridiculous to me, and they probably do to you as well. When these types of situations arise for you, it's important that you gently allay the concerns and barbs from friends and help them understand that knowing more about producing your own food is not weird—it's a basic element of life.

Gardening produced its share of surprises as well, including a near-poisoning when Jenna tried to eat her first potato. "I picked these potatoes and they were all green on the outside, but I started cooking them anyway. I had this gardening cookbook beside me and I read this little sidebar and it basically said, 'By the way, green potatoes can make you very ill.' And I was about to eat a huge plate of home fries!" Whew! Good thing Jenna likes to read.

Jenna says her immersion in farm life resulted in lessons every day. Fortunately, they were not always major disasters, though she did have a few mishaps with livestock that you can read about in her own book. "There were all these little revelations," she recalls. "As a total beginner, you just don't know anything!"

After about a year, Jenna felt the itch to move again, but not back to the city. She headed to Sandgate, Vermont, leaving behind the rugged Rockies for the verdant shadows of the Green Mountains. "One of the reasons I took a job in Vermont was that it allowed me to be rural," she explains. Jenna now rents a cabin on six acres where sheep, rabbits, her two Siberian Huskies, and a host of fowl (chickens, geese, and ducks) roam. "This year I even raised my own Thanksgiving turkey," she boasts. Jenna also keeps bees, which she began to learn how to do in Idaho.

## Becoming Aware of Food and Farming

The single greatest lifestyle change I usually observe among folks who are new to food production and hobby farming is an increased awareness of and a greater appreciation for their natural surroundings. As a lover of the natural, rural world, this is very encouraging to me. What's interesting is that along with that awakening comes a certain responsibility, and for many, a strong desire to pass that knowledge along to family, friends, and the general public. It seems this is the impetus that often leads to starting a small farm business or a food-making hobby.

Another natural result of a new awareness is simple: Real food tastes better. Now then, what is "real food"? When I say "real food," someone usually tosses out, "Then what is fake food?" I've run into variations of that one many times because the name of my meat business has the word *natural* in it. At a farmers market, there is invariably a smart aleck who comes

## How to Live Like a Local, Part One

More and more urban folks are moving to the country. But they are finding that rural life is much different than city living. Some try and change the rural area to fit their urban expectations, while others try and fit into their new community and adopt the local culture and customs. To help this latter group, the University of Missouri Extension system has developed a training course on rural living: Rural Living 101, a two-week crash course for ex-urbanites offered in Kirksville, St. Joseph, and Nevada. Class members learned about battling noxious plants, pesticide laws, stocking ponds, land use management, and more. The Vernon County clerk tutored participants about the township form of government, where a board meeting can take place at an elected clerk's home as often as at a public building. While these are all important things to know, I think they are missing a few essential elements of adjusting to rural life. So I would like to begin a two-part series that offers a few tips for the would-be rural resident preparing for a life in the boonies.

One of the first things you need to learn is the rural address system. No, not the official one that has things like CR 400 east and 200 north, which is used only by the post office and the volunteer fire department. The real address system, used by the locals, involves such addresses as "The old Wilson place." This is the farm that was settled by Jack Wilson at the turn of the century but has not been in the Wilson family since 1965. There are also landmark locations like

up and says, "If your meat is natural, then is what I've been eating for 40 years unnatural?" Depending on the person's typical cuisine (boxed dinners filled with preservatives and ingredients that are unpronounceable), the answer may be yes!

To me, real food is about taste, texture, care, and production practices. Real food is unaltered—food you can pick from a garden and eat without much transportation. Real food is food you prepare yourself with a knowledge of where it came from and how it was produced. This definition is not from a textbook; real food can be defined in a lot of ways, I think. But it is basically the awareness that real food comes from a farm somewhere, no matter how small, and not from the grocery store or drive-up window.

Whatever real food means to you, use this awareness for good. Maybe your awareness includes starting a hobby farm, or maybe it means getting involved locally and supporting organizations that mean something to you. Learn and share—that's a good start!

"About a mile past the place where the road curves at the old oak tree." The old oak tree is actually an old oak stump, because the tree was taken out by the tornado of '72. Distances are different in rural areas. "A mile or so" means at least three miles. "It is not far at all" means it will take about 45 minutes to get there. "You can't miss it" means bring your cell phone and call when you get lost.

There are different driving rules in rural areas. You can always tell city folks on the road because they drive on the left or right side. Locals drive down the middle. City folks might notice there are no speed limit signs in rural areas. This is because the locals do not believe in speed limits. There are also very few street signs at intersections (see rural address system above). If you are following a pick-up and you see someone mowing a ditch bank, be prepared to stop. The pick-up in front of you will be stopping and will likely have at least a five-minute conversation with the person on the mower. Speaking of machines on the road, do not try to pass slow moving farm equipment on the road. It is dangerous for you and the tractor driver and will almost guarantee you are not spoken to at the next church picnic. Speaking of courtesy, don't flip off the old guy in the pick-up that cut you off; there are loaded guns in his gun rack, and he is a good shot.

(By Gary Truitt. Reprinted with permission from *Hoosier Ag Today*, November 30, 2008.)

## First-Year Tips

One thing I've noticed about a lot of business books is that they provide resources, tips, and forms for a business start-up, but very little advice about how to survive the first year of your new venture. I am willing to bet this is due to the fact that a lot of business books are written by professionals, executives, and professors and not by people who have actually started, with their very own money and very own backaches, a new company. That's why I wrote this section.

I've started two small businesses. One is my meat company, Aubrey's Natural Meats, and the other is my communications and writing firm, Prosperity Ag Resources and Communications. I haven't seen all the scenarios that might come your way, but I've weathered my share of good and bad choices—many of which you've already read about in previous chapters.

## Cold Antler Farm

Life in Vermont is, so far, all that Jenna hoped it would be. Since moving there, she has named her little place Cold Antler Farm, and she's a regular blogger about progress, process, and peacefulness around her little homestead. Though Jenna says she'd like to get married and have a family out there, for now her animal companions, particularly her Huskies, Annie and Jazz, keep her comfy and warm.

"I always thought I'd do this 'farm thing' when I had a family or as a retiree," Jenna admits. But apparently, the call of the wild was just too strong to ignore. Calling the Huskies her "other car," Jenna enjoys hooking up the dogsled and touring the area on snowy winter days. The sled is six feet long with long runners, and she describes it as looking like a chair on skids. A long line attaches to the dogs' collars. She even takes a sled ride the half mile to fetch her mail.

The farm is close to her office, making it generally convenient. "It's a nice 11-mile drive to work. I come back at lunch and get eggs, feed the sheep, let the dogs out, and then head back in an hour. So far, I can make it work!" she says proudly.

Jenna enjoys her garden and bakes bread almost every weekend. She's also taken up the fiddle and found a group of friends who like to play instruments, too. Jenna is loving her life, but always dreaming about the next step. Her goal is to save money and acquire her own place in the future. Sometimes the slow route to her goals gets frustrating and seems far away. "The hard part of hobby farming is the reality of mortgages, banks, limitations like no credit, and not having inherited a farm," she says. "No matter how much you want something, sometimes you just can't have it right away. It's hard to be a hobby farmer and want to be a full-time farmer."

Jenna's on her way, but for now she keeps busy and tries to strike a balance between her two worlds. "I'm not quite there yet; I'm feeding chickens in the morning and then going to the office."

She continues, "Right now the number one crop on this farm is my blog. The goal is within ten years to be a full-time farmer and writer. I'm slowly crawling up that hill."

## Success Factors for First-Year Survival

There are several key factors to consider when you are thinking about how to survive the first year. These important factors for your new business are:

- Financial preparedness
- Manageability of your business design
- Pacing and planning your progress
- Conscious learning
- Organization

These five factors are the specific areas of your new business life that I've found can easily cause a problem if neglected or handled poorly. In my first year in business, I did well in some areas and really fell down in others. The ones I missed are on the list today because I realized after the 13th or 14th month that they were causing problems only because they weren't fixed.

I'd like to offer you some suggestions for getting your first year going strong without wanting to either quit or cry. Okay, you still might want to cry, but I don't want you to quit. Wipe away those tears of frustration and move on. You don't have to follow these suggestions exactly, but I advise you to examine each of the six areas and at least make sure you're dealing with each topic.

Soon enough, you'll know the names and even the brands of all the farm machinery. This is Robert from Ravens Glenn on a blue New Holland harvester.

## Living Like a Local, Part Two

In Part One, I discussed how city folks who move to a rural community can adjust to their new environment. I discussed how to get around using the unofficial rural address system and the proper etiquette for driving on rural roads. I was inspired to write this series by the Missouri Cooperative Extension system, which has developed a "Rural 101" course to teach newcomers to rural areas the facts of rural life. While very informative, as most Extension programs are, it lacked some important basics, like how to communicate with your neighbors. Interpersonal communications is the subject of Part Two of this series.

There are some important phrases you might hear from your new neighbors that may sound strange. Here is a translation. "The fair" means the county fair, which you must plan to attend. "The game" means the Friday night basketball game at the local high school. Plan on attending even if you don't have kids in school. Next to Sunday church services, it is the most important social event of the week. "Slow time/fast time" in certain parts of Indiana refers to Eastern Standard Time (fast time) and Central Standard Time (slow time). "Town" is the location of the nearest bar, hardware store, coffee shop, or auto supply shop. Size is not relevant here; the town may be 10 or 10,000 people. "County office" may refer to either the Cooperative Extension office or the USDA office. It depends whether you are filing crop insurance papers or 4H project papers. Be aware that the "Young Farmers" organization has farmers over 40 years old and the "Extension Homemakers" are, in most cases, not young mothers but grandmothers and great-grandmothers. You

# Financial Preparedness

Being prepared financially will bring you the most benefits when you're making your plan to survive the first year. If you are financially prepared and are a good financial manager (or are married to one or hire one), your life as a hobbyist or entrepreneur will not only be a great deal easier, but also has the possibility of lasting long term.

## Wean Off Other Sources of Income

The single smartest thing I did before I started to work for myself full time was to plan financially for the upcoming changes. I began more than a year should also know that "all-y'all" is proper speech when you live within sight of the Ohio River.

As a new rural resident, there are some essential skills you must acquire. You must be able to tell the difference between the smell of cow, pig, and chicken manure. You should learn the names and colors of the most popular types of farm equipment: John Deere is green, New Holland is blue, Case IH is red, and so on. You would also be wise to learn to tell the difference between a planter and a combine; this will save you from acute social embarrassment.

If you plan to live in a primarily agricultural area, you need to learn the proper seasons of the year: planting (March–May), haying (June–August), harvest (September–November), and the meeting season (December–February). There are also some weather terms you should know. "Rain up your way" means: did it rain within 10 miles of your home. "Cold ain't it" means: did the temperature drop below -10 degrees (anything warmer is not worth mentioning). "Hot ain't it" means the temperature went above 110 degrees (anything less is not worth mentioning).

Now, if you follow all these suggestions, you will make some of the best friends you have ever met. You will experience a sense of community that is rapidly disappearing today. You may be the recipient of incredible generosity and outstanding kindness. And, if you stay in the rural area long enough—say ten years or so—you will still be referred to as "the new folks in town."

(By Gary Truitt. Reprinted with permission from *Hoosier Ag Today*, December 7, 2008.)

in advance by slowly weaning myself off outside income. I knew I wanted to quit my town job. I knew I wanted to live off the land and off my intellectual abilities. I knew I didn't want a boss anymore. I also knew I was going to have a lot less money.

When you realize that less money will be coming in for the foreseeable future, you've got to get serious about planning and be honest with yourself and your family about what you need and don't need. Spending less on unneeded items in the early days of a new venture is easier than making enough money to cover what you spend on them. I started with services and began to systematically delete them. I cancelled my once-a-month housekeeper, stopped taking so many clothes to the dry cleaner, started washing and detailing my own vehicles (okay, my husband did those things, but it was still free), canceled my wine clubs, and laid off the lawn boy. I even started coloring my own hair and buying hair-care products from the drug-store, not the salon (I know there are some women out there who just gasped). Hey, I never said all the choices were going to be fun.

Once I'd removed some reoccurring expenses that weren't necessary, I began to look at how I was using my money. I took out a certain amount of cash every week, and my husband and I used it on incidentals such as eating out. That was it for the week, and we stuck to it. The simple little things you can do to control your finances include making a household budget, which should demonstrate areas where you can lean up and things you can live without.

I don't want you to feel deprived, although learning to spend less to prepare for making less initially may seem like going on a diet. If your business is successful, though, you'll soon be back at the day spa. Now I've added back several of the things I canceled when I first started. For example, I now have a twice-a-month housekeeper, a lawn boy, and an intern. Oh, and I rehired my hairdresser—thank heavens! Some of the practices I implemented have stuck with me, though: I still take out the week's cash, and I buy fewer dry-clean-only clothes.

## Pay Down Debts—Even Good Debts

We hear a lot about credit card debt and the devastating burden too much unsecured debt can have on a person or a family. Yet even good debt—that is, debt tied to assets such as a car—can be a burden when you're spending more money or earning less. My advice is to pay off your car or student loans or debt for big-ticket electronics while you're in the process of con-sidering starting a new business venture. Doing so really will help out. I paid off our pickup truck (and kept it for eight years before we traded it in) and sold an extra vehicle for cash to pay off that last couple of thousand on my student loan. We also refinanced the house for a lower rate and a lower monthly payment.

How much more comfortable would you be if you didn't have a $700-a-month car payment? Would it make you more likely to pursue your dream of a small farm or a side business? If bills are standing in your way, pay them off first, or consider selling the asset if you can. You'll have enough stress working the new venture without worrying about your bills at the same time. Do this for yourself, your family, and your sanity.

Jenna has a full-time job off the farm, but every day still begins and ends with livestock chores.

# Manageability of Your Business Design

Dr. Phil's catchphrase "How's that working for you?" is exactly what I mean by manageability. When you get started, do your best to select selling venues, products, services, and a general design for your business based on the manageability of the model for you and your family. Many small farm ventures fail because the business model is inconvenient. Don't allow this to happen as you get started.

What is more difficult to predict is the parts of the business that start out manageable and turn into nightmares—or at least into bad dreams. I've had my share of these, including my short bout with a retail store. You'll have to decide for yourself whether something is worth it or not, but decide based in part on convenience and your ability to manage it within the

## The Rural Life

Jenna says reasons to move to the farm can be as varied as the individual, though she doesn't think of herself as a trendsetter. "I don't think I'm unique at all. People now are raised in an overtechnological world. We're the first generation to be able to have all this instant gratification. It's almost numbing now," Jenna says of the lack of awe-inspiring experiences that many young people have. "I felt so excited when I ate the first salad that I grew myself."

She continues, "There is a relief to be away from all this technology. Not that farming doesn't have technology—it does—but it's *outside!*"

"I think farming appeals to people of a certain demeanor, whether they lived 300 years ago or will live 300 years in the future. Most people aren't interested in farming full time, but they can do other things in town, like have an urban garden." For example, Jenna says a good friend who lives in downtown Cincinnati grows the best arugula she's ever tasted.

For Jenna, the farming lifestyle is what she wants—all of it. "I'd rather spin my own yarn and then learn to knit with it than be buying it at the store." She believes that people of all ages are becoming interested in self-sufficiency as a result of feeling less safe. She also thinks people increasingly want to be prepared for disasters.

For some, though, farming may be about capturing the lost entrepreneurial spirit of our ancestors. She says, "For an independent generation, I think farming is just the ultimate rebellion!"

confines of your day. Set up or change your business to work for you, not the other way around.

The other function of manageability is financial. Can you afford to do what you're doing? Is this a business model that will make enough money to cover your expenses, or will it eventually turn a tidy profit? What do you want from it—profit as a business or enough cash flow to consider it a large-scale hobby? Work the business to your desired result.

# Pacing and Planning Your Progress

The Divine Hand of Progress—isn't that what we all want to be blessed by? Actually, I think it depends on whether you're starting a business or a hobby, on how busy you want to be, and on how much money you'd like this venture to make. My best advice is to resist the temptation to expand too quickly. If you're doing any kind of sales with your new venture, then eventually progress will come to you—for many, sooner rather than later.

This is one of those first-year survival tips that would have helped me. We went from having our beef in one restaurant the first week in business to 25 restaurants by the third month. By the eighth month, we were down to ten. Is that progress or regress? You could call it either, but the truth was that we were growing too big too fast. It goes back to that manageability thing—we just couldn't handle all that business so quickly. The infrastructure wasn't there for deliveries, and neither was the cash flow for raw materials. The lesson was expensive and exasperating both for me and for some of my customers.

While a "full" product line is great for space in the retailer's refrigerator case, you're better off adding products gradually as you find success and customer loyalty with your first few items. Become special to your customers with a few things they love, and when you add products, do so by building excitement rather than trepidation. It is better to surprise and excite customers with new products in your own time than to disappoint them when you can't deliver on a promise.

Another key to pacing your progress is managing spending. Keep the money you've accumulated for start-up under close watch, and set a threshold or reserve. When you begin to see yourself inching (or tumbling) toward that reserve, immediately evaluate what you're spending the money on. Are you going too fast? Are you dipping into your reserve to deal with a short-term issue, or do you have no idea when you'll be able to replace it? How do you feel about your progress?

Finally, take time to plan your next steps by keeping notes on ideas and potential new products and venues as you go. This step often helps budding entrepreneurs feel more comfortable about slowing down without fear that they'll forget the great idea they had. When I first started my meat business, if a great inspiration came to me I wanted to implement it immediately, giving my husband and employees another ball to juggle. I learned (again, the hard way) that this is often a mistake. Take the idea and file it someplace where you're sure you'll be reminded of it later. Or ponder the idea for a while, fleshing it out on paper during a moment of down time or in conversations with people you trust.

Now, when I get a major brainwave I put it on the calendar in my task list with a date assigned to it. That way, in, say, a month, the idea pops back up and I can reevaluate it when I'm a little more reasonable and a little less inspired.

## Conscious Learning

During your first year, you'll be learning constantly, and most of that learning will be through immersion. You may think you don't need any more learning, but in fact, the opposite is true. Take time to keep growing and learning in your chosen area. One major reason is to stay current. Things are always changing.

For some people, when they start a new venture they become consumed by it and do not stop to observe what's going on around them. This, too, is a mistake. What if you're working hard to implement something new and you don't realize that the farm down the road has been doing it successfully for two years? What if you're struggling to get through an issue, but if you'd just gone to the vegetable growers' conference last month you could have heard a professional address that topic and then discussed it with peers?

Get involved and stay involved. When you are in business for yourself, knowledge is power and ambition is the engine. Don't forget to take time to learn away from your daily work.

## Get Organized!

Neatness is a virtue! Neatness also makes life easier and, in the case of business, less expensive and less frustrating. Start a filing and billing system *before* you've sold one single item. Order the software and enter the dates and accounts at the same time as you are buying labels and deciding on delivery routes. You need directions to get to a new market, and you need labels to identify your product. Likewise, you need a way to track sales and manage your taxes. Don't overlook the importance of getting organized early and doing so often.

If you forget to pay a bill because it gets stacked under a crate, late fees are wasted money and bad for your credit. Likewise, if you get too busy to bill your customers, you won't get paid. Don't get stuck wondering why you don't have any money even though you've been doing lots of selling, simply because you've failed to collect!

Organization also applies to other parts of your new venture. Set up an inventory system and a system for storage and filing. Create easy access to vendor contact information for your supplies and equipment and phone numbers for help and support. If organization is just not "your thing," hire someone to do it for you weekly. The money spent will be well worth it if you would otherwise let invoices and mail stack up. In just a few hours, you could save days of potential migraines.

## *Making the Transition*

Though it is obvious Jenna loves living in the country, she has endured her share of transition woes. Life is different out here, after all, even if it is only a few miles to town. Jenna noticed that garbage pickup was not available in her area, nor was cable television or high-speed Internet access. "The sacrifices of conveniences you make were shocking to me; not debilitating, but shocking— and some are still shocking!" Jenna admits. "Sounds like you're trying to live a simpler life, but it is not always that way."

During her few years as a country girl, trial and error played a big part in Jenna's learning curve. "I wish sometimes that I had an ag degree," she begins the tale of her first garden—a self-proclaimed disaster. "As a designer, I wanted it to look a certain way. I planted it for aesthetics, not for cropping and agriculture."

Jenna's tolerance for bad weather has increased, and she's adjusting to making sure the animals are cared for on inclement days. "You have to get used to power outages that last days, not hours. I've gotten used to storing water, too. There will be a lot of little things that you're just not used to."

Although she has spent several years learning about animals and producing her own food, she still views her life as one in transition. Her advice: Take your time and don't overwhelm yourself the first year. "I started with five hens and one hive of bees, and with the most basic, simple steps with each thing," she says. "You can build from there."

Jenna also believes that those small steps can begin in any situation, regardless of region, zip code, or housing scenario. "Anyone in Manhattan can begin by learning to can tomato sauce. Even if you live in Topeka, Kansas, in an apartment, you can grow tomatoes in a pot." Food and farm are, in part, a mindset, and Jenna thinks you've got to train the mind and, just perhaps, the body will follow. "Keep the lifestyle in mind and move toward it."

No matter the hardships, Jenna has a clear goal in mind and a love as deep as a valley for small farm life. With farming, there are some harsh realities, but she believes it is worth it. "I'm hoping that it remains, at its heart, a lifestyle. I'm hoping that because it will be so hard to get there, there will always be reverence."

## You Want to Do WHAT?

"I'm the only person in my family who wants to farm," Jenna says with a little laugh. Like everyone in America, it seems, Jenna knows that a few generations back *someone* in the family tree farmed. But no one in the recent past, that's for sure. She figures it has been at least three or four generations since a member of her family lived on a farm.

When Jenna's mom found out her daughter planned to become a shepherd, her reaction wasn't overwhelming joy. "Mom was like, 'Oh, that's great! Four years of college for this?' She would definitely be more comfortable with me in Manhattan than on a farm in Vermont," Jenna says. "To her, there is safety in numbers, not in isolation. Mom has asked me flat out why I want to do this. But I think people definitely find fulfillment in many different types of things."

Her family has worried about everything from burglars to workload, but Jenna waves them away without much concern. As far as burglars go, her two massive Huskies are a rather obvious deterrent. And as for farm labor, Jenna likes it. "The labor is the easy part," she says. She's done all sorts of labor around her place, including daily feeding of her livestock and hauling water. She's also become handier after having to build structures and fences.

"Part of growing for me was realizing that I didn't have to do the same thing as my parents or have the same lifestyle I grew up with," Jenna explains simply. "I'm happiest when I am out on a farm."

## Jenna's Best Practices

- You've just got to do it.
- Learn and read and study.
- Don't overwhelm yourself; start small and grow.
- Learn to do stuff with your hands and live with less technology every day.
- Have patience with your transition.

# Finale

# The Prints You Leave Behind

## Finding Your Stride and Yourself

One aspect of this journey that I think you'll be pleasantly surprised with is the chance to learn something new about yourself and discover a part of life you didn't know existed. No matter what your age or station in life, when you consider starting a new business, it's akin to asking yourself what you want to be when you grow up. If you've never owned your own company, you'll find out a whole lot about yourself and your ability to handle problems, deal with people, and adapt. If you're an entrepreneur who's considering a move to the country from the city, I have to tell you, you're also going find out you who you were before "the move" and learn who you've become.

What is it like to change your lifestyle? For each person it is unique. For some, it is exciting; you may have to force yourself to use restraint as you approach critical decisions. For others, the lifestyle change is intriguing but intimidating at the same time.

This has been a business book about the rural and local food lifestyle, but there will be choices and distractions and decisions where no other person's advice could ever help. To me, that's the beauty of the consumer's journey to becoming a producer. Go with your instincts. In your walk on this earth, let your footprints on the land leave small marks but deep impressions. Let this journey allow you to be who you want to be.

Best wishes from one small farmer, now, to another.

# Resources

## By Sarah Beth Aubrey

Aubrey, Sarah Beth. *Starting and Running Your Own Small Farm Business.* Storey Publishing, 2007.

Prosperity Ag Resources and Communications
*www.prosperityagresources.com*

Sarah's Blog
*http://prosperityagresources.wordpress.com/*

## Profiles

**Aubrey's Natural Meats**
Sarah and Cary Aubrey
8570 N. 750 W.
Elwood, IN 46036
*www.aubreysnaturalmeats.com*

**Jeannie Ralston**
220 North Zapata Highway 11
Laredo, TX 78043
*www.jeannieralston.com*

**Salem Road Farms**
Brent and Suzie Marcum
340 South Salem Rd.
Liberty, IN 47353
*www.salemroadfarms.com*

**Sherwood Acres Beef**
Jon Bednarski
3001 Ballard School Rd.
LaGrange, KY 40031
*www.sherwoodacresbeef.com*

**Loleta Cheese Company**
Bob Laffranchi
P.O. Box 607
Loleta, CA 95551
*www.loletacheese.com*

**Lost River Market & Deli**
Brad Alstrom
26 Liberty St.
Paoli, IN 47454
*www.lostrivercoop.com*

**Cook's Bison Ranch**
Peter and Erica Cook
5645 East 600 South
Wolcottville, IN 46795
*www.cooksbisonranch.com*

**Flying Tomato Farms**
Rebecca M. Terk
117 Forest Ave.
Vermillion, SD 57069
*http://flyingtomato.wordpress.com*

**Baarda Farms Veggie Van**
Denise Baarda
1566 River Rd.
Mt. Bethel, PA 18343
*yaunky@epix.ent*

**Shuckman's Fish Company and Smokery, Inc.**
Lewis Shuckman
3001 West Main St.
Louisville, KY 40212
*www.kysmokedfish.com*

**Raven's Glenn Winery**
Renee Guilliams
56183 County Rd. 143
West Lafayette, OH 43845
*www.ravensglenn.com*

**Cold Antler Farm**
Jenna Woginrich
70 Sophies Way
Arlington, VT 05250-9576
*www.coldantlerfarm.blogspot.com*

# Books

Adams, Barbara Berst. *The New Agritourism: Hosting Community and Tourists on Your Farm.* New World Publishing, 2008.

Ralston, Jennie. *The Unlikely Lavender Queen: A Memoir of Unexpected Blossoming.* Broadway Books, 2008.

Woginrich, Jenna. *Made From Scratch: Discovering the Pleasures of a Handmade Life.* Storey Publishing, 2008.

# DVD

**Shared Wisdom: Selling Your Best at Farmers Markets**
This DVD provides helpful techniques for selling at a farmers market. Also includes interviews, PowerPoint presentations, and printable documents.

# *Agriculture and Trade Groups*

**Agricultural Innovation Center**

*www.fire.rutgers.edu*

Based at Rutgers University, offers resources and assistance in business development.

**Agricultural Marketing Resource Center**

*www.agmrc.org*

This site is specifically aimed at producers creating value-added agricultural and food products and is a major resource for data, research, and information on studies done in all aspects of agriculture and agribusiness.

**Belted Galloway Society**

*www.beltie.org*

Society for Belted Galloway breeders that works to promote the breed. The web site offers breed information, including breed history, guidelines for selection, and breed data.

**Community Supported Agriculture/Robyn Van En Center**

*www.csacenter.org*

Based at Wilson College, this web site provides resources and information about community supported agriculture.

**Ecological Farming Association (Eco-Farm)**

*www.eco-farm.org*

This organization offers seminars, resources, and training for a variety of sustainable agriculture and food projects.

**Food Co-op 500**

*www.foodcoop500.org*

A resource that supports the development of food cooperatives around the nation. The site includes a 74-page manual about how to start a cooperative venture.

**Food Routes**

*www.foodroutes.com*

An organization that supports the development of local food systems and encourages buying local.

**Local Harvest**

*www.localharvest.org*

Provides a search feature for local CSAs and an online community with forums, blogs, and a newsletter.

**Midwest Organic and Sustainable Education Service (MOSES)**

*www.mosesorganic.org*

An education and resource service for farmers of all sizes looking to learn about or transition into organic food production.

**National Cooperative Business Association**

*www.ncba.coop/index.cfm*

A national group for cooperatives. Offers education, cooperative development, and political representation of cooperatives.

**Organic Trade Association**

*www.ota.com*

This membership organization represents organic products of all kinds (products and goods, not just food) in U.S. business and also collects and distributes statics on organics.

**Purdue University's New Ventures Team**

*www.ces.purdue.edu/new*

This team of professionals and professors provides resources, online business planning tools, and a wealth of advice and information for start-up businesses in food and agriculture.

**Slow Food USA**

*www.slowfoodusa.org*

This organization represents the idea of food and a community of life and is a worldwide movement.

# U.S. Government Resources

**Country of Origin Labeling (COOL)**

*www.countryoforiginlabel.com*

Web site describes the newly implemented COOL regulations and program rules.

**Department of Energy (DOE)**

*www.energy.gov*

Provides information on the agency's programs, grants, and initiatives.

### Environmental Protection Agency (EPA)
*www.epa.gov*
Provides information on the agency's programs, grants, and initiatives.

### Sustainable Agriculture Research and Education Program (SARE)
*www.sare.org*
This is a federal program that offers grants and education nationwide for sustainable agricultural practices.

### U.S. Department of Health and Human Services
*www.grants.gov*
The most comprehensive listing of federal funding for grants.

### U.S. Fish and Wildlife Service (FWS)
*www.fws.gov*
Information about wildlife, including articles, grants, permits, and regulations.

## Small Business Administration

### SBA
*www.sba.gov; www.sba.gov/localresources*
The Small Business Planner section has information about writing a business plan. The Services section has information about financing, special audiences (including women and veterans), and online training programs. The Tools section offers a library and resources and forms.

### Small Business Innovation Research Program (SBIR)
*www.sba.gov/aboutsba/sbaprograms/sbir/index.html*
A grant and contract program for eleven federal agencies that award money for research and innovation in a variety of businesses.

## USDA

### Agricultural Marketing Service (AMS)
*www.ams.usda.gov*
Provides information and resources about helpful AMS programs, focused on marketing successfully in the age of technology.

### Community Supported Agriculture
*www.nal.usda.gov/afsic/pubs/csa/csa.shtml*
Provides general information and resources about CSAs.

### Conservation Reserve Program (CRP)
*http://www.nrcs.usda.gov/programs/CRP/*
Provides information about the CRP, which offers farmers financial and technical assistance focused on conserving resources.

### Cooperative Extension Service
*www.csrees.usda.gov/extension/*
Search for your local cooperative extension here.

### Environmental Quality Incentives Program (EQIP)
*www.nrcs.usda.gov/PROGRAMS/EQIP/*
Provides information about EQIP, a voluntary program that promotes the relationship between agricultural production and environmental quality.

### Farmers Market Search
*www.apps.ams.usda.gov/FamersMarkets*
National directory of farmers markets since 1994.

### Renewable Energy for America Program (REAP)
*www.rurdev.usda.gov/rbs/farmbill*
This program offers guaranteed loans and grants to farmers and rural small businesses each year to implement energy-efficiency measures or create renewable energy.

### Rural Business Cooperative Service (RBCS)
*www.rurdev.usda.gov/rbs*
Offers information about rural development, cooperative programs, and rural businesses.

### Value-Added Producer Grant (VAPG)
*www.rurdev.usda.gov/rbs/coops/vadg.htm*
An annual grant program for agricultural producers looking to add value to a raw product and create a value-added product for sale.

# Media Resources

### Farmers Market Today
*www.farmersmarkettoday.com*
Magazine offers resources and articles to help small farmers become more profitable and successful.

### Highway 6: Your Road to the Country
*http://insightadvertising.typepad.com/hwy_6_your_road_to_the_co/*
Michael Libbie hosts this radio show that provides news and stories for rural and farm populations.

### Hobby Farms
*www.hobbyfarms.com*
Magazine provides articles about small farms and living in the country. Also includes recipes, equipment and livestock information, and recreational activities.

### Hoosier Ag Today
*www.hoosieragtoday.com*
Gary Truitt hosts this radio program focused on Indiana agriculture. Provides news topics, market prices, and interviews.

### MaryJanes Farm
*www.maryjanesfarm.com*
Magazine written by an organic farmer that provides advice, stories, and an online community focused on women farmers.

### Reiman Publications
*www.reimanpub.com*
Site provides access to many rural-focused magazines, including *Country Living, Country Woman,* and *Farm and Ranch Living.*

# Other Web Resources

### AllBusiness.com
*www.allbusiness.com*
This web site offers comprehensive business tools (such as business plans and business advice) at no charge. It's a good resource for business plan templates.

### Cultivating the Web
*www.eatwellguide.org*
Free PDF publication provides farmers, foodies, and markets with ideas for using online resources that can be used to promote and grow a business.

### My Own Business
*www.myownbusiness.org*
An online course in starting your own business. The site includes good resources for business plan templates.

### WordPress
*www.wordpress.com*
The service I use that offers free blogs.

# Sample Business Documents

## Business Plan

Starting on the next page is a sample business plan for Penny's Herbs & More, a fictitious herb business in the Midwest. This is one example of how to develop a business plan; it may need to be expanded for your own business. In the Financial Details section, the numbers were intentionally left out. Costs will be individual for each business and in each part of the country. This section was included to show you the types of expenses and costs that might be incurred.

# Penny's Herbs & More

## Business Plan

### I. Mission Statement

Penny's Herbs & More will offer the freshest, highest quality culinary and medicinal herbs to local residents seeking flavor and health enhancement.

### II. Executive Summary

Penny's Herbs & More is a small business created to produce and sell a high quality herb selection for the local area. We provide a supply of choice herbs to individual buyers through farmers markets. Our products are naturally raised and we are working to become USDA Certified Organic. We believe the natural and organic movement in foodstuffs is growing and will continue to gain in mainstream acceptance. Our customers can proudly say the herbs they buy are from plants that are:

- Raised in a private garden in the Midwest, currently in Indiana.
- Natural, with no pesticides, in a garden that is in organic transition.
- Clean and safe to eat.

Besides the highest standards being applied in growing these herbs, our customers can enjoy the personal touch of quality and service. Our customers are buying:

- Select herbs such as basil, chives, dill, garlic, oregano, parsley, rosemary, tarragon, thyme, chervil, lemon balm, and mint.
- Food-enhancing quality and medicinal value that is impeccable and much more consistent than store-bought brands. Because we are involved in every aspect of production, from planting, to growing, to harvesting, to cutting and cleaning the herbs, we can personally ensure the integrity of our product.
- Custom cut and bundled herbs without additional cost. Our clients specify which herbs they would prefer to ensure their individual use.
- One contact for delivery and sales.

### III. Background

The company's founder, Penny Lane, has been involved in the herb gardening business since her childhood days of hobby gardening with her mother. Along with her sister, Rose Thorn, she continues today to plant and grow select herbs. On the botanical side, Mrs. Lane serves as a judge for the 4-H program of Indiana, judging gardening and horticulture projects. She is also a member of the Master Gardener's Association and has been a part of the Master Gardener Association's Garden Walk. Further, along with her degree in Biology, she has a minor in Botany from Purdue University. Since college, she has focused primarily on hobby gardening and herb growing. From 1999 to 2001, Mrs. Lane worked in greenhouse strategic planning with plant and flower companies, and since 2001 she has been managing her own herb garden.

Ms. Thorn also possesses much relevant experience. She is the biology teacher for Sunny Days High School, where she has her students participate in nature walks and outside leaf and flower projects. She also judges numerous horticulture and gardening 4-H projects each year. Thus, she is highly competent in the biological and growing features intended for high-quality herbs. Prior to teaching, Ms. Thorn spent 10 years being self-employed in the flower and gardening business. Finally, Ms. Thorn has retailing and sales experience from her employment at a small, family-owned greenhouse.

## IV. Products Offered
We will offer our fresh herbs through farmers markets. Penny's Herbs & More may be visited at two local farmer's markets on Thursday evenings from 4:00p.m.-7:00p.m. in the town of Jasmine, Indiana, and on Saturday mornings from 8:00a.m.-12:00p.m. in the town of Honeysuckle, Indiana. Our products will also be available for mail order online. More specifically:
- High-quality sun herbs:
    a. Basil, chives, dill, oregano, rosemary, tarragon, thyme
    b. Gift baskets and special orders
- High-quality shade herbs:
    a. Chervil, lemon balm, mint
    b. Gift baskets and special orders

## V. Target Clientele and Target Market Niches
Our customer is the average, day-to-day consumer. Largely, our target population is a 25-65-year-old, middle class to affluent female, either single or with a family. We expect her to be interested in the product for these reasons: health/diet, food safety, culinary use, medicinal value, and because natural foods are trend-setting.

In addition to this, our end customer is also likely to be a middle-upper class to affluent male, ages 30-65, who pays attention to herbs for the enhancement of grilling meats and the peace of mind of knowing that the food is accented with wholesome herbs to sanctify a delicious grilled entrée.

## VI. Short-Term Goals and Objectives
The numbers below are an estimate of customers for start-up:
- 10-15 regular consumers buying 5-10 bundles of prime herbs weekly.
- Approximately 100 bundles/week start-up. Total bundles of product moved per week, start-up: 1,500.
- Grow to 5 total farmer's markets in first 12-18 months.
- Add seasonal flowers and vegetables, by summer season 2010, if needed, or any time during years 1-2.
- Become profitable so that a driver for deliveries can be hired by September 2010.
- Increase sales by 20% by end of second year.

## VII. Long-Term Goals and Objectives
- Expand sales in retail to our own line of markets in Indiana.
- Experience sales growth of at least 20% each year through building supply relationships.
- Add, at retail, exotic plants and herbs, any time we see a demand.

## VIII. Financial Details

### Start-up Expenses

1. Equipment
   a. Seeds and plants $_____
   b. Gardening tools (pots, shovels, shears, pruners) $_____
   c. Watering aids (hoses, attachments, timers) $_____
   d. Plant labels, markers, tags $_____
   e. Plant supports $_____
   f. Pest deterrents $_____

   Equipment total: $_____

2. Fertilizer
   a. Organic fertilizers $_____
   b. Organic potting mix $_____
   c. Mulch $_____

   Fertilizer total: $_____

3. Transportation Costs (during start-up)
   a. Fuel $_____
   b. Maintenance $_____

   Transportation total: $_____

4. Marketing Materials $_____

   Marketing total: $_____

5. Licensing and State/Federal Requirements
   a. Board of Health $_____
   b. Merchant's retail license $_____

   Licensing total: $_____

6. Legal/Accounting/Consultative
   a. Set up LLC or incorporate $_____
   b. Review contracts $_____

   Legal, etc., total: $_____

7. Insurance
   a. Liability $_____
   b. Worker's compensation $_____

   Insurance total: $_____

**Grand total start-up expenses: $_____**

### Projected First-Year Sales and Revenues

Moving a projected 1,500 bundles of herbs through farmers markets and some Internet sales.

| | |
|---|---|
| Total annual sales | $_____ |
| Annual projected pre-tax income | $_____ |
| Minus expenses | $_____ |
| Annual projected revenue | $_____ |
| Pre-tax earnings retained at 30% | $_____ |

Payout to debt on business, to be done quarterly $_____

Projected pre-tax salary expenses to owners $_____

Projected employee expenses $_____
(2 at $_____ $10/hour; 40 hours/week; 30 weeks/year)

Annual dollars retained for taxes, to be paid quarterly $_____

# Grant Application

## Kentucky Proud POP Grant Application

Farm/Company Name: <u>SHERWOOD ACRES BEEF</u> PH<u>502-222-4326</u>

Contact Name: <u>Jon Bednarski</u> PH: <u>502-222-4326</u>

Farm/Mail and E-mail Addresses: <u>info@sherwoodacresbeef.com</u>

Taxpayer ID Number : <u>61-1377363</u>
*(required by state to issue grant checks)*

How much do you propose spending on behalf of Kentucky Proud? <u>$4,890.14</u>

How much are you requesting as a match (up to 50 percent)? <u>$2,445.07 (50%)</u>

1. Please document the total expenses or services that you are claiming are relevant to Kentucky Proud. Eligible expenses include printing, labeling or packaging costs related to use of the Kentucky Proud logo, UPC codes, demos, food samples, point-of-sale materials, advertising or other expenses relevant to promoting or advertising a retail food product. Include copies of actual receipts or price quotes. *Use separate sheet if necessary.*

2. Please specify how your product affects Kentucky family farms. Be specific if your ingredients come directly from such examples. This is an important criteria for deciding grant acceptance. *Use separate sheet if necessary.*

3. Please describe how visible this project will be to consumers for brand awareness. For example, you might note that the Kentucky Proud logo will be printed on the front of your product's package and just as large as your company name, etc. *Use separate sheet if necessary.*

4. Please describe the dollar value or economic impact of this project. *Use separate sheet if necessary.*

5. Include details about your business and product. Do you use a certified kitchen? Do you have food liability insurance? Do you have proof of compliance of meeting health and food safety standards with your products? Is this item already offered at a supermarket or restaurant? *Use separate sheet if necessary.*

**Please mail this completed application and all attachments to: Roger Snell, Kentucky Department of Agriculture, 100 Fair Oaks, 5th Floor, Frankfort, KY 40601.**

### Supplement to Kentucky Proud POP Grant Application

Numbered responses below correspond with same numbers of questions on grant application.

1) The amounts below reflect our plans for the 6 month period of March-August. Bids, documentation or previous invoices for same items are attached.

- **$275.00** Booth space fee to promote Sherwood Acres Beef at "Oldham County Showcase" (Product and services show by the Oldham County Chamber of Commerce) March 6,7,8.
- **$300.00** Membership in Kentucky Farm Bureau "Roadside Market" program
- **$100.00** Membership fee Oldham County Farmer's Market
- **$150.00** Membership fee "Heart of St. Matthews" Farmer's Market
- **$1,416.96** Advertisements in "Oldham Era" newspaper to promote new retail store (3" advertisement @ $59.04 x 4 insertions monthly x 6 months) Ky pround logo in ad
- **$468.00** Advertisements in "Roundabout" newspaper to promote new retail store (monthly display ad @ $90 per insertion x 6) KY proud logo in ad
- **$375.00** Upgrades to web site (currently @ 6 pages. Adding 1 page for retail store promotion @ $225 + text revisions to existing 6 pages @ $25/page) ky proud logo on pages
- **$649.00** Retail store signage. KY proud logo on sign (two separate bids attached – one for sign, the other for sign frame)
- **$418.70** Graphics wrap for retail store freezers. KY proud logo on wrap
- **$464.00** Upgrades to Farmer's Market displays (2 new table skirts with graphics, 2 new vertical banners for existing sign frames) KY proud logo on skirts and banners
- **$273.48** Ongoing labeling costs. (3,000 labels) KY proud logo on labels

TOTAL:**$4,890.14**

2) Our beef is raised on our Kentucky family farm (Lagrange)

3) Kentucky Proud logo appears on all of our materials (signs, brochures, labels, web site, advertisements, and Farmer's Market displays)

4) 2009 goal is to increase sales volume by 50% from 2008. To achieve this we will need

5) We are a Lagrange, KY beef "Farm to Table" operation. We have liability insurance and proof of health and safety standards compliance. Our beef is USDA inspected. Our beef is currently available at small retail markets and restaurants as well as being offered direct to consumers at Farmer's Markets, through phone orders, and soon to be available at our own retail store.

# Co-op Application

*Lost River Community Cooperative*

## MEMBERSHIP APPLICATION

Name _____ (legal member of record)

Street Address: _____

City: _____ State _____ Zip: _____

Phone (home/work/mobile) _____ Phone (home/work/mobile) _____

E-mail Address: _____

### TERMS AND CONDITIONS

* I agree that only persons living in my household will use this membership.

* I certify that I am at least 18 years of age.

* I understand that the "Legal Member of Record" is the person to whom all official co-op mailings are addressed and to whom official voting rights accrue in all co-op elections.

* I understand that full rights of membership are granted upon full payment of the membership fee.

* I understand that as a member I am agreeing to support the mission and goals of the co-op and to abide by the provisions of the Articles of Incorporation, the Bylaws and policies of Lost River Community Co-op as they now exist or may from time to time be amended.

* I understand that this application for membership is subject to the approval of the Board of Directors and that my membership is subject to the Articles of Incorporation, the Bylaws and Policies of the Lost River Community Co-op.

* I agree to pay a one-time lifetime membership investment of $90.00. I understand that this is refundable upon my terminating membership in good standing.

Signature: _____ Date: _____

Mail to: Lost River Community Co-op
112 W. Water St.
P.O. Box 505
Paoli, IN 47454

Sign up for our yahoo group discussions: http://groups.yahoo.com/group/orangeco-op/
Contact Steven Spurgeon for any questions: strahbale@yahoo.com

MEMBERSHIP NUMBER _____          AMOUNT PAID _____

# Marketing Plan

This is a sample marketing plan for A Cut Above, a retail meat market featuring local, superior products. It comes from the University of Tennessee Center for Profitable Agriculture.

**A Cut Above** — Summary of Planned Promotional Activities

|            | Year #1 | Year #2 | Year #3 |
|------------|---------|---------|---------|
| Month #1   | Grand opening celebration Local feature articles on the farm page of one newspaper Activate 1-800 number | Begin frequent buyer program Direct mail campaign | Sampling campaign |
| Month #2   | Daily radio ads on two stations & two print ads per week in one newspaper for the month | Local feature articles | Local feature articles |
| Month #3   | In-store promotions with point-ofpurchase displays | Local recipe contest | Host annual festival |
| Month #4   | Host Chamber Coffee | Host Chamber Coffee | Host Chamber Coffee |
| Month #5   | Target local media for free publicity by submitting news leads | Target local media for free publicity | Target local media for free publicity |
| Month #6   | Implement a sampling campaign at retail store | Sampling campaign | Sampling campaign |
| Month #7   | Host media day | Host media day | Host media day |
| Month #8   | Coupon campaign through newspaper, radio, direct mail & flyers | Coupon campaign | Coupon campaign |
| Month #9   | Unveil web page and distribute magnets & pencils with printed web site address | Barbeque promotions | Barbeque promotions |
| Month #10  | Barbeque promotions | Pork promotions | Pork promotions |
| Month #11  | Pork promotions | Host Open House | Host Open House |
| Month #12  | Seasonal features | Seasonal features | Seasonal features |

**A Cut Above** — Estimated Marketing Budget (Three Years)

|  | Year #1 | Year #2 | Year #3 |
|---|---|---|---|
| Initial Marketing (Includes market research, surveying and initial advertising & promotion tools) | $9,500 | $3,000 | $3,000 |
| Month #1 | $1,200 | $80 | $200 |
| Month #2 | $800 | $100 | $100 |
| Month #3 | $400 | $350 | $1,000 |
| Month #4 | $350 | $350 | $350 |
| Month #5 | $80 | $80 | $80 |
| Month #6 | $150 | $200 | $200 |
| Month #7 | $300 | $300 | $300 |
| Month #8 | $600 | $500 | $500 |
| Month #9 | $1,200 | $500 | $500 |
| Month #10 | $500 | $500 | $500 |
| Month #11 | $500 | $500 | $500 |
| Month #12 | $200 | $200 | $200 |
| **Total Budget** | **$15,780** | **$6,660** | **$7,430** |

# *Press Release*

## *Penny's Herbs & More*

**FOR IMMEDIATE RELEASE**

**CONTACT:**
Penny Produce
Penny's Herbs & More
Phone: (222) 555–1212
Fax: (222) 555–3434
E-mail: penny@fresherbs.com
Web site: www.fresherbs.com

### Penny's Herbs & More Announces New Market Location

Penny's Herbs & More is expanding to offer an additional market location this spring. Opening on May 1st, the new location will be in Anytown, PA, at the popular Anytown Farmer's Market. Look for Booth Space 11 and the familiar Penny's Herbs sign in front of our signature green tent.

This location will carry the same product line as the current farmers market in Newtown, where Penny's Herbs & More has been selling products for two years. Owner Penny Produce will be on hand each market morning from 8 until 10 a.m. to greet customers. The new location has been requested many times by customers and will provide shoppers with a chance to buy Penny's products in two communities.

Markets are open starting May 1st from 8 to noon each Saturday. The market closes for the season October 1st.

For additional information on the Anytown Farmers Market, visit www.anytownmkt.org. Penny's Herbs & More serves the local area with fresh, naturally grown herbs and assorted greens grown on Penny's own small farm.

– END –

# Newsletter

# Raven's Glenn Winery Cellar Talk

www.ravensglenn.com                                      Vol. 4, Issue 4

**Contact Info:**

Raven's Glenn Winery
56183 CR 143
West Lafayette, OH 43845

740-545-1000

crg@ravensglenn.com

Sharing your e-mail address with us is a privilege and for this we are grateful.

Our policy is to not allow others to have access to our customers' e-mail information.

## PRE-HOLIDAY RELEASE OF GOLD MEDAL ICE WINE

Fall is in the air and with the onset of cooler weather it's time for this year's release of the 2007 Raven's Glenn Gold Medal Ice Wine.

Among the several gold medals won for this year's release, we are particularly proud that we are one of only a small handful of wineries whose ice wine qualified for the Ohio Quality Wine program gold seal offered by the Ohio Grape Industries Committee.

As always there is a very limited quantity of this rare and very special dessert wine. Our 2007 ice wine may be purchased at the winey or via our online wine store at $25.99 per 375 ml bottle. A truly excellent value.

## HOLIDAY GIFT SET NOW AVAILABLE

To our knowledge, Raven's Glenn is the only winery in the Midwest to offer a unique holiday gift set. It's a perfect gift for friends and relatives or, impress your customers or employees by giving this special Raven's Glenn holiday set.

Each ready-to-give set includes your choice of either our Raven Rouge semi-sweet red wine or our very popular White Merlot semisweet blush wine, two monogrammed Raven's Glenn wine glasses, and an embossed Raven's Glenn cork screw all in a silver gift box ready for holiday giving.

Call the winey at 740-545-1000 for special orders or alternate wine selections including special pricing for volume purchases.

# THANKSGIVING DAY BUFFET

Consider Raven's Glenn for your Thanksgiving Holiday dinner and leave the cooking to us. Last year's feast was a sell-out. Make your reservations soon. This year Chef Mike has put together plans for a lavish buffet filled with the traditions for the holiday.

Thanksgiving Day Buffet
$18.95
Children 10 & under $10.95
Includes nonalcoholic beverage
Served from 11am till 3pm
Tasting bar open 11am till 4pm
Please call 740-545-1000 for reservations

# RAVEN'S GLENN ONLINE STORE

For a complete selection of our wines, you can shop online at www.ravensglenn.com for those hard to find wines only available at the winery.

# RAVEN'S GLENN GIFT CERTIFICATES

Raven's Glenn Gift Certificates are available online at the Wine Store or by telephone at the winery at 740-545-1000

# ENJOY A SUNDAY CHAMPAGNE BRUNCH ON THE RIVER

Our Sunday Champagne Brunch continues to grow in popularity and at $18.95 it's a great value. (Children 10 and under $10.95) Reservations are not required but it is always a good idea to make that reservation. Brunch is served from 11:30am till 2:30pm. Tasting room will be open Sundays from 11am till 4pm for wine sales.

# LIVE ENTERTAINMENT THURSDAYS & FRIDAYS

Dan Barnes, the magical master of the grand piano performs every Thursday and Friday evening from 5:30 to 8:30. Dan is incredibly talented and a really fun guy. Join us for a wonderful evening of fun, music, wine, and good food. Make your reservations now for a fun evening.

### Restaurant Hours:
Tuesday thru Saturday 11am till 8pm (last seating)
Live entertainment 5:30 till 8:30 on Thursday and Friday
Sunday Champagne Brunch 11:30am till 2:30pm

### Tasting Room & Gift Shop Hours:
Sunday 11am till 4pm
Monday 11am till 6pm
Tuesday thru Saturday 11am till 8pm

# A GENTLE REMINDER

With the approaching holiday season... it's a good idea to call in advance for restaurant reservations, especially on weekends and for the Sunday champagne brunch.
CALL 740-545-1000

Restaurant specials are changed daily, but prime rib is only available on Wednesday evening.
Visit our website for menu selections
http://www.ravensglenn.com

Sincerely, Bob & Renee Guilliams
Raven's Glenn
Ohio's Crown Jewel of Wineries

Raven's Glenn Winery is operated by Raven's Glenn Ltd

# Index

# About the Author

Sarah Beth Aubrey was born into a family of multigenerational farmers and was raised to believe that, like Scarlett O'Hara in *Gone with the Wind,* it is land that always sustains. After receiving an agriculture communications degree from the University of Illinois, she was never satisfied working off the farm. She returned to production agriculture and the promotion of home-raised products with the founding of her company, Aubrey's Natural Meats, LLC, in 2003.

Sarah is a member of her state Farm Bureau organization, serving as county woman leader, and is a graduate of the Indiana AgriInstitute, the state's premier agricultural leadership development organization. She has served in an advisory role to the Indiana Board of Health, with a special emphasis on farmers market regulations.

Sarah is the owner of Prosperity Communications, a writing, speaking, and consulting practice focused on rural development and growth in the areas of value-added agriculture, renewable energy, and energy efficiency. She offers grant writing assistance as well. Her book *Starting and Running Your Own Small Farm Business* was released in January 2008. A collection of Sarah's writing can be found on her blog, http://prosperityagresources.wordpress.com. The blog focuses on rural business development and features articles, excerpts, useful links, and advice for finding funding opportunities.

She and her husband, Cary, live in rural Indiana and raise beef cattle.

## Photo Credits

Sarah Beth Aubrey: pages vii, 8, 19, 35, 38, 43, 62, 67, 82, 89, 90, 91, 97, 99, 105, 126, 150, 200, 228, 230, 256–258, 264; Robb Kendrick: page 4; Ralph Lauer: page 7; Ravens Glenn Winery: pages 15, 201, 205, 212, 218, 237, 265–268; Steve Rawls: page 21; Suzie Marcum: pages 26, 31, 34; Jon Bednarski: pages 46, 51, 259, 260; Sylvia Bednarski: page 53; Amy Reohr and Cindy Davy: pages 68, 69, 81, 84; Cook's Bison Ranch: pages 108, 111, 115, 121; H. L. Scholten: pages 130, 137, 143; Rebecca M. Terk: page 145; Sabrina Baarda: page 151; Nate Gould: pages 154, 160; Denise Baarda: page 168; Kara Keeton, Keeton Communications: pages 175, 178, 183, 193, 197; Tricia Weill: page 229; Joanna Chattman: page 241; Lost River Community Cooperative: page 261; University of Tennessee Center for Profitable Agriculture: pages 262, 263; Cary Aubrey: page 275.